# NUCLEAR PAKISTAN

# NUCLEAR PAKISTAN

DR SHALINI CHAWLA

*Introduction by*
Air Commodore **Jasjit Singh** AVSM, VrC, VM (Retd)

KNOWLEDGE WORLD
KW Publishers Pvt Ltd
New Delhi

*in association with*

Centre for Air Power Studies
New Delhi

2011 FPBA BEST PUBLISHERS AWARD (ENGLISH)

First reprint April 2013

The Centre for Air Power Studies is an independent, non-profit, academic research institution established in 2002 under a registered Trust to undertake and promote policy-related research, study and discussion on the trends and developments in defence and military issues, especially air power and the aerospace arena, for civil and military purposes. Its publications seek to expand and deepen the understanding of defence, military power, air power and aerospace issues without necessarily reflecting the views of any institution or individuals except those of the authors.

Jasjit Singh
Director General
Centre for Air Power Studies
P-284, Arjan Path
Subroto Park
New Delhi 110010

Tele: (91-11) 25699131
E-mail: diroffice@aerospaceindia.org



Published in India by
Kalpana Shukla
KW Publishers Pvt Ltd
4676/21, First Floor, Ansari Road, Daryaganj, New Delhi 110002
T: +91.11.23263498 / 43528107 E: knowledgeworld@vsnl.net

ISBN 13: 978-93-81904-25-1

Printed and bound in India

*For my son Arush*

*with the fond hope that the nuclear weapons*

*threat would cease to exist during his lifetime……*

# Contents

# PREFACE

For more than two decades, Pakistan has relied on nuclear weapons to conduct its grand strategy (of indirect approach) against India. The central assumption on which Pakistan's nuclear weapon programme has been built and progressed has been that a credible nuclear deterrent would compensate for the inferiority of its defence forces. Pakistan's defeat in the 1971 War led the defence planners to believe that the nuclear weapon was the only option to deal with India. In this respect, Pakistan adopted a doctrine and strategy not very different from that pursued by the North Atlantic Treaty Organisation (NATO) against the USSR. Pakistan visualised nuclear weapons as the sole guarantor of its national pride and national survival, and, thus, bilaterally started to seek Chinese assistance for its nuclear weapon programme in the late 1960s in the aftermath of the US sanctions. The build-up of the weapon programme was done clandestinely, seeking the European and Chinese technical assistance.

Nuclear weapons for Pakistan were also seen as a projection of Pakistan's scientific and technological capabilities. In addition, Pakistan has perceived nuclear weapons capability as a "currency of power" which would enhance its strategic image in the international community, in general, and the Muslim countries, in particular. In this respect, Pakistan was able to gather financial, political and psychological support from the wealthy oil rich Muslim countries for building the *Islamic bomb*. Given this background, Pakistan's export of nuclear technology to the Muslim world – Iran (Iraq probably turned down the offer), Libya and, probably, Saudi Arabia, was no surprise.

Pakistan perceived its nuclear weapons as providing a foolproof guarantee of its sovereignty and survivability. The central logic around the build-up of nuclear weapons has been to neutralise India's conventional capability and make it unusable in a scenario of a conventional confrontation. With this basic assumption, the declaration of the "first use" nuclear doctrine by Pakistan was also no surprise.

Pakistan, till today, after fourteen years of overt nuclearisation, does not have a written doctrine because it believes that an unwritten doctrine would help in maintaining robust deterrence. However, the policy-makers have, from time to time, talked about the basic tenets of a doctrine which include: Indo-centricity, credible minimum deterrence and first use. Ironically, while the doctrine has intentionally been kept ambiguous, Pakistan has been extremely vocal and clear about its command and control system and also, to a large extent, about its delivery systems, no doubt, to enhance its credibility and legitimacy.

The objective of Pakistan's nuclear weapons is to deter any form of Indian military aggression which could potentially occur in response to state sponsored terrorism. Pakistan has deliberately adopted a posture of irrationality and the Pakistani leadership is convinced that they have managed to deter India with this posture of irrationality and uncertainty, more than once in the past. The central strategic assumption of Pakistan's nuclear strategy has been that India cannot impose a conventional war on Pakistan, leave alone achieve a decisive victory, without the risk of catastrophic consequences of a Pakistani nuclear attack. The Pakistani leadership believes that Pakistan's possession of a nuclear arsenal and its first use policy would be sufficient to deter war. This logic in Pakistan has been reinforced by the common perception that it was able to deter Indian military action on various occasions even after highly provocative terrorist attacks, including the one in November 2008.

Pakistan has continued to expand its nuclear arsenal, along with the delivery systems, including ballistic missiles, cruise missiles and combat aircraft. The delivery systems have been growing primarily with the Chinese support, despite Pakistan's financial constraints. Pakistan has consistently claimed that it has been forced into the missile build-up given India's growing missile muscle, but the reality is that India's strategic requirements are far more complex and demanding compared to Pakistan's. India needs to take the existential China threat into consideration in its strategic and defence planning, which is not the case for Pakistan. Pakistan's belief in deterrence and increasing the options of delivery systems provide it with enhanced deterrence. For Pakistan, the ballistic missile capability enhances its deterrence, and also its targeting choices for the preemptive strikes, given its "first use" nuclear doctrine. Also, wide options have opened up, ranging from the tactical nuclear weapons (e.g. the NASR missile) and long range ballistic missiles (the Ghauri and Shaheen-2).

Presently, Pakistan has the fastest-growing stockpile in the world, and it is estimated that in 2011, it possessed a nuclear weapons stockpile of 90-110 nuclear warheads, which is a steep rise compared to what Pakistan possessed even three years ago in 2009, when the number of warheads was estimated at 70-90. Pakistan's unwritten doctrine, with a heavy reliance on the first use posture, accompanied with Pakistan's domestic instability, due to the raging insurgency, certainly makes the study of its nuclear arsenal, strategy and its future, critically important for the Indian policy-makers, defence planners and all Indians interested in national security issues. This study has been undertaken with a hope of creating a better understanding of Pakistan's nuclear strategy from an Indian perspective. The study makes an attempt to evaluate the future of the rapidly expanding nuclear arsenal in one of the most troubled nations. The security of Pakistan's nuclear weapons is a global concern currently, but India would have to bear the direct implications in the event of any breakdown in the safety and security of the weapons.

I believe any book is an outcome of direct and indirect consultative effort and it is almost impossible to name everybody who has contributed in this study. My ideas and thoughts were shaped during various discussions and meetings over a period of two years. There are many people who have helped me in this study and I extend my heartfelt thanks to all of them. Some obviously deserve special mention. First and foremost, I would like to thank my mentor, Air Cmde Jasjit Singh (Retd), Director General, Centre for Air Power Studies (CAPS), who has constantly encouraged and supported me. I am deeply indebted to him for spending his precious time in discussions with me which have largely helped me structure my thoughts and complete this study.

All my colleagues at CAPS have been of enormous support, and I am specifically thankful to Dr Manpreet Sethi and Lt Col. Rajeev Ghose for their valuable responses to bouncing off my ideas and providing a critical view-point at various stages which helped me to continuously improve my study.

I am thankful to Shri Jayadeva Ranade, Distinguished Fellow, CAPS, for helping me to understand China better and provide a deeper insight into the Pakistan-China strategic relationship. Air Mshl T.M. Asthana, generously shared his extremely useful experience during his tenure as the C-in-C of the Strategic Forces Command. I would like to extend my gratitude to Brig Gurmeet Kanwal for happily giving his invaluable advice whenever I approached him.

I am grateful to the library staff of CAPS, specially, Mrs Sushma and Ms Viji who have coped up with my never ending requests for books and journals with a smile. The pantry staff of CAPS helped to keep up my energy levels with the constant supply of refreshing beverages, which are badly needed for a study of this nature. I sincerely thank Ms Rehana Mishra, for her unstinting meticulous help in editing this book. I would also like to thank Jose Mathew, Anoop Kamath and the entire team of KW Publishers for their cooperation and perseverance. The shortfalls in the book remain my responsibility.

This book could not have been completed without the support of my husband, Vikas, who takes immense pride in what I do and has been extremely patient and supportive of my work. His honest feedback on my writings and ideas has been the biggest contributor in my growth. My son Arush, obviously, deserves special thanks for being so understanding and happily allowing his mother to be in her study, shutting him out for long hours. I am truly blessed to have loving parents and extremely caring parents-in-laws who have always encouraged me in all my endeavours. Their unstinted support has given me immense comfort and allowed me to work with a completely carefree attitude. My brother-in-law Vishal and sister-in-law Chhavi have been of great help in myriad ways. My brothers Amitesh, Abhishek and Bhupesh and my sister-in-laws Nitika and Krunal have always been a tremendous source of strength during this project.

# Pakistan's Nuclear "Deterrent?": An Introductory Essay

Jasjit Singh

The roots of Pakistan's nuclear strategy and deterrent lie in the concepts and ideology of the state itself and the system of governance it has adopted. The military has ruled Pakistan directly and indirectly for nearly five decades. Unlike the period till the 1970s, the role of the bureaucracy has declined. Within the larger military dominance of Pakistan's national governance (or misgovernance) and security policies, it is the Pakistan Army which has exercised overwhelming influence and control. And nowhere was this so visible than in the question of command and control of nuclear weapons when the army insisted after the May 1998 nuclear tests that the post of Chairman, Joint Chiefs of Staff, who would presumably be the seniormost military officer to advise the Prime Minister on strategic and nuclear matters, till then held in rotation, should be manned only by an army officer!

**Dominance of the Army in the Power Structure**

It is necessary to recall that national security is defined by the core values and vital interests of a state. At the time of the creation of Pakistan, its founding father, Mohamed Ali Jinnah had outlined a set of core values that essentially visualised equality among the people, with religion remaining a personal and social phenomenon with which the state was not involved and which would not interfere in the affairs of the state.[1] Pakistan, however, moved away from these core values early on in its history and started to

Air Commodore **Jasjit Singh** AVSM VrC VM, IAF (Retd), awarded the Padma Bhushan for a life-time's contribution to national security and defence, is the Director General, Centre for Air Power Studies, New Delhi.

establish core values based on the argument for partition, that is, all human beings are not equal and Muslims constitute a different nation, and, hence, Pakistan! This, in turn, created an ideological conflict with India which laid down its core values in its Constitution, basing them on the belief that all human beings are equal (and, hence, liberty, equality, social justice, etc.) By the end of the first three decades of its existence, Pakistan had come close to establishing Islam as the basis of the state itself. An added factor shaping the security policy was the centrality of the military (essentially the army) in the power structure of Pakistan.

Hasan Askari Rizvi, in a landmark study of the military in Pakistan's politics has identified five factors responsible for the increasing participation of the army in non-military governance of the state:

- The sense of insecurity of the new state following the partition riots and demographic shifts where the military moved with determination to provide stability;
- The weak social base and disarray of the political leadership and erosion of political institutions;
- Discipline and cohesion of the military (and bureaucracy);
- Utility of skills for civil purposes; and
- The high prestige of the military as a competent, impartial, and effective organisation.

The deterioration of conditions inside Pakistan gave rise to "two closely related trends in Pakistan. First, the calls to the Army to help the civil government in the maintenance of law and order gradually increased. Second, the military became an important factor in the decision-making process of Pakistan."

The creation of Pakistan on the basis of religion also led to its army having to "adapt to Islamic principles and practices."[2] This in itself started to alter the secular, apolitical moorings inherited from the (British) Indian Army. Those responsible for training and doctrine insisted that their traditions go well beyond the inheritance derived from the Indian Army under the British. In reality, this implies a search to establish medieval foundations for the history and traditions of the army, rather than a modern, secular, professional basis for them. Medieval Muslim armies were the king-makers, and rulers emerged mostly out of them; and they, in turn, relied on the field commanders for kingship. The result, in the judgement of one scholar, was, "There are armies that guard their nation's borders, there are those that are concerned with protecting their own position in society, and there are those that defend a cause or an idea. The Pakistan Army does all three."[3]

Hostility towards India has its roots in ideology, but it has also been fostered in Pakistan by the dominant groups, especially the bureaucratic-military-feudal complex that has ruled Pakistan for more than six decades. The army can claim a special responsibility and status for national survival as the defender against the "Indian threat", and the bureaucracy has sought to project the idea that a strong, stable administration (which, according to it, requires a central role for the bureaucracy, and the "corrupt" politicians to be kept at a distance) is needed for the same reason. Christina Lamb made a very perceptive analysis when she wrote, "While the Pakistan Army has a casting vote, it is hard to see how relations between Pakistan and India can ever be peaceful, as it is in their interest to maintain high tension to justify the large proportion of the budget spent on defence (twenty times the education expenditure), particularly amid declining US interest after the Soviet departure from Afghanistan."[4]

There is no transparency in Pakistan's security policy. The high degree of opaqueness is symbolised by the fact that even 65 years after independence its defence budget indicates only the total figure of funds allocated and senior military leaders have publicly acknowledged that this does not include the procurement of major weapon systems.[5] The nuclear weapons capability has emerged from a clandestine programme aided and abetted by external powers, some of who, over time and depending upon their own perceived national and strategic interests, have altered their policies related to this programme. There is, thus, scanty information available about Pakistan's nuclear weapons capability leave alone its likely strategy. And yet the issue is extremely important. An attempt, therefore, has been made in this study to assess Pakistan's nuclear strategy based on available information and deductive analysis of circumstantial evidence.

**Strategic Aims and Objectives**

Few countries in the world clearly articulate their strategic aims and the developing countries maintain an even more ambiguous position on this aspect. For a country like Pakistan which does not allow any transparency in its military posture, the challenge of trying to accurately assess its strategic aims becomes even more difficult. The best that can be done is to analyse the writings and statements of senior military leaders and the broader strategic community and try to correlate them with actual events and developments. Looking at the available literature, we find that a former head of Pakistan's premier strategic think-tank outlined six main strategic aims and objectives of Pakistan in the late 1990s which provide us with important clues in understanding the likely nuclear strategy.[6] These are discussed in the following paragraphs.

***"To Strengthen National Power"***

This would be a reasonable goal applicable to all countries. National power also encompasses multiple dimensions ranging from the economic, social, political and military to the intellectual. The issue in all cases is not the absolute secular aim of strengthening national power but what becomes critical for peace and stability is how that power, especially its military component, would be used.

***"To Prevent Open Aggression"***

Prevention of aggression against the territorial integrity and sovereignty of the state is an unexceptionable aim and one which is an article of faith with all sovereign states. Aggression itself is not easily definable, especially when it gets linked with irregular and unconventional warfare. But what is curious here is the term "open" for which no apparent explanation has been forthcoming. The question that arises is whether this is only a generalised aim and a rhetorical statement or carries deeper meaning in terms of politico-military strategy? The weight of historical evidence points to the fact that all wars in the past were initiated by Pakistan, including the start of the war in 1971 through coordinated air attacks on Indian installations.

***"To Induce India to Modify its Goals, Strategies, Tactics and Operations"***

This is a highly subjective formulation and depends upon Pakistani perceptions of India's goals, strategies, etc., including operations. At the broader strategic level, this aim would appear to seek to influence, and intervene in, not only the internal affairs of India but also to change the social, economic and political basis on which India as a state is established. In other words, Pakistan's strategic goals aim to question and alter the core values on which India has established itself, thereby threatening India's security. With nearly 14 percent of the population in India being Muslims, Pakistan hopes to influence Indian ideology to get it harmonised with its own which does not recognise the innate equality of human beings. Islamisation of Pakistan under Gen Zia ul-Haq's military regime intensified the ideological conflict between India and Pakistan. Promotion of political Islam into India with the use of terror and violence forms a key component of Pakistan's pursuit of its ideological goals.

***"Attain a Position of Security or, if Possible, Dominance, Which Would Enhance the Role of Other (Non-Military) Means of Conflict"***

This finally gave a clear indication of the nuclear strategy as the Pakistani elite believed it [rather than any government commitment to it though circumstantially this was followed by the army and Inter-Services Intelligence (ISI)]. The strategy would provide a "dominant" security umbrella under which Pakistan could pursue terrorism as a tool of state policy against India. The first part of the goal is rational and would be seen as part of the national strategic aims of all countries. But the second part is revealing. Pakistan not only seeks "dominance" over India in security terms but the purpose of this dominance has also been clearly articulated. Such dominance is to be sought not necessarily only for application of military force, but to allow greater potential for the application of "other (non-military) means of conflict." We are not told what these means of conflict would be. But it is reasonable to assume that they would encompass the use of violence, potentially through terrorism and militancy, engineered or supported by Pakistan. This is borne out by other indicators.

The Pakistani military has professed to change its military doctrine from one of "active defence" to that of "offensive defence" after 1971. This was not clearly articulated except to some extent by the then Army Chief Gen Aslam Beg during the "Zarb-e-Momin" military exercise in 1989 and its rationalisation by those (like Mushahid Hussain) who appear to have received deep background briefings. But then, Gen Beg very soon linked it to the military nuclear deterrent. In conventional terms, the doctrine was applied almost every time, and "Pakistan has not hesitated to be the first to employ the heavy use of force in order to gain an initial advantage."[7]

As regards the "non-military" means, Pakistan has made it amply clear that this implies covert war through guerrilla warfare and *jihadi* terrorism. Brig S. K. Malik has interpreted the Holy Quran in search for guidelines, and has placed a strong emphasis on this (mis)interpretation that the Holy Quran enjoins upon the believers to use terror as a weapon of war. Such an interpretation goes totally contrary to the teachings of Islam. But the military in Pakistan interpreted it to suit their own strategic and ideological goals to conclude, "The Quranic military strategy, thus, enjoins us *to prepare ourselves for war to the utmost in order to strike terror into the hearts of the enemies, known or hidden...*" (emphasis in original).[8] Gen Zia-ul Haq, the Army Chief and the Chief Martial Law Administrator at that time, commended the book in its foreword. The book teaches that terror must be struck during the preparatory stage, in the run-up to war, during war, and for war termination. "Terror struck into the hearts of the enemies is not only a means, it is the end in itself.[9] Once a condition of terror into the opponent's heart is obtained, hardly anything is left to be achieved. It is the point where the means and the ends meet and merge."[10] Gen Beg and other army leaders almost inevitably cited the central thesis of the book to exhort military officers when addressing them, especially at the Quetta Staff College and National Defence College.

Three dimensions of the use of terror deserve note. Firstly, by seeking the authority of the Holy Quran, it seeks to deny any opportunity for dissent in the Islamic Republic. Secondly, Pakistan has always complained of a threat from India because of the latter's much larger size and potential, The major *means* of imposing terror over a professed but superior enemy would be through the *use and threat of use* of nuclear weapons since they remain weapons of mass destruction and mass terror. Since 1990, in fact, Pakistani leaders have been increasingly referring to the threat of nuclear weapons, and even believe that this has deterred India from attacking Pakistan in retribution for its proxy war in Punjab and Jammu and Kashmir (J&K) in India. Thirdly, terror can be spread through terrorism. The strategic concept also lays great stress on the psychological effects of the use of terror as a weapon. This has provided the conceptual base and legitimacy (of the Holy Quran) to prosecute terrorism in India at tragically high levels since the early 1980s.

A related aspect is the role of guerrilla warfare in the army's strategic thought. For a conventional professional army brought up on British military traditions, it has surprisingly placed significantly high emphasis on guerrilla warfare. It has been argued that Pakistan enjoys an advantage over India in this mode of warfare because it is more cohesive, and ideologically is more highly motivated. This may well be the reason for the place of honour for guerrilla warfare in Pakistan's strategic doctrine. The initial interest in guerrilla warfare may have been generated by the invasion of J&K in 1947. But the close association of the army with the US military, starting from 1952, gave it a strong impetus. A Special Forces unit was set up in 1959 which grew into the SSG (Special Services Group) where the previous Army Chief and President, Gen Pervez Musharraf served two tenures.

But there has been a fundamental difference in the approach of the USA and Pakistan. The American doctrine was primarily concerned about *suppressing* such a war, but the Pakistanis also studied it in terms of *launching* a "people's war against India."[11] The Pakistan Army's contacts with the Chinese Army added strength to the Pakistani version of the People's War doctrine, with the difference that China's doctrine was strategically defensive while Pakistan developed it with a strategically offensive concept. It is significant, therefore, to note that, in 1989, the then Army Chief, Gen Aslam Beg had blamed the earlier defeats (of 1947-48, 1965, and 1971) on "lack of strategic vision." He had gone on to assert that Pakistan had a clear strategic vision now. One aspect of this was undoubtedly the launching of the militant proxy war in J&K from July 31, 1988; and the other was the achievement of nuclear deterrence (to provide 'defence' in the offensive defence strategic doctrine, while irregular war was used for the 'offensive' component) by 1987.[12]

***"To Promote and Capitalise on Advances in Technology in Order to Reach Parity or Superiority in Military Power"***

The search for *parity* with India has been a psychological obsession with the Pakistani elite from the time Jinnah described the new state as "moth-eaten." Given the asymmetry of capabilities between the two states, a central pillar of Pakistan's strategy from the very beginning was to acquire additional strength through external linkages to achieve parity and, where possible, even acquire military superiority. On the face of it, this appears unrealistic and we have tended to dismiss it as rhetoric. But a close examination would indicate that this was a conscious policy that did provide superiority even if it was fragmented.

***"To Deter War"***

Given the asymmetry in size and capabilities, it is near impossible for Pakistan to deter India from launching a war. If India has not done so, it is essentially because of India's choice as a national policy and strategic culture which believes that force should be used only as a last resort. But nuclear weapons can play a key role in achieving this aim. It is significant that unlike the position followed by India, that its aim is to prevent war and nuclear weapons are meant to deter a nuclear weapons threat only and not war itself, Pakistan has sought to deter war which, if it takes place, would almost inevitably be a conventional war, at least at the beginning. Pakistan has often asserted that under these circumstances, nuclear weapons would be employed first by it at an early stage, with a high risk of escalation to a nuclear exchange.

Deterrence of war could be achieved through denial or punishment or a combination of the two. Since Pakistan recognises that India maintains significant conventional military superiority, the denial dimension of deterrence can only rest on assumptions of superior generalship and fighting capabilities. The earlier wars against India were launched on the assumption that the Pakistan military can always defeat the superior numbers of the

Indian military; but the assumption failed every time, even as late as in 1999 on the heights of the Kargil sector when Pakistan possessed declared nuclear weapons, and even held out a threat of nuclear weapons, but failed to deter India from using its Air Force since Pakistan's threat of nuclear weapons was not interpreted as credible (except to impress Washington, which also failed). Pakistan's record in making wrong assumptions is history.[13]

**Nuclear Policy Sans a Doctrine**

In my view, Pakistan has a nuclear policy of sorts, but without any doctrine of substance, and, hence, the ambiguity and changing stances in the nuclear policy and its articulation. Pakistan was far too preoccupied with its clandestine programme to acquire nuclear weapons to pay much attention to how they would be used and to what effect. An almost juvenile simplistic belief prevailed right up to the actual acquisition of nuclear weapons (and even after) that this would neutralise India and provide Pakistan with a parity status and freedom of action in using force against India. The more perceptive commentators had started to argue, without giving any rationale, that nuclear weapons would enhance Pakistan's security. For example, its very intelligent and capable Foreign Minister for a long time, Agha Shahi, who as late as 1996 claimed that Pakistan's nuclear weapons were grounded in its security calculus to deal with what he termed as the "Sword of Damocles" of an irredressible conventional military balance hanging over Pakistan.[14] This logic, which seeks to make a credible case for the acquisition of nuclear weapons by Pakistan on the lines of nuclear weapons in and for the North Atlantic Treaty Organisation (NATO), also debunks Pakistan's diplomatic stance of the so-called initiatives and posture [like the South Asian Nuclear Weapon Free Zone (SANWFZ), position on the nuclear Non-Proliferation Treaty (NPT) and Comprehensive Test Ban Treaty (CTBT), etc.]. More important from our point of view, it does not indicate anything about the doctrine and strategy beyond the other statements that Pakistan's strategy would be based on the concept of first use of nuclear weapons.

Hence, the first publicly known articulation by Lt. Gen. Khalid Kidwai after the nuclear tests had been carried out was of the oft quoted "four spaces" defining the Pakistani nuclear threshold. These were expressed to two Italian professors in November 2000 who had little knowledge of nuclear weapons and less expertise on nuclear strategy. In fact, the most credible articulation was to come from a troika of senior retired officials of Pakistan: the acerbic Agha Shahi (former Foreign Minister), Air Chief Marshal Zulfiqar Khan, and Abdul Sattar (former Foreign Minister), who wrote a joint article advocating a deterrent doctrine of "first use in the last resort."[15] But till now, no official has even obliquely endorsed this.

A perceptive Pakistani commentator had stated that Pakistan "has to copy the qualities outlined in the Indian doctrine."[16] Pakistan, therefore, started to project its nuclear doctrine as one of a minimum deterrent but one which is country specific and, hence, perceived to be "more manageable." There is also recognition that having achieved a nuclear status quo, conventional military build-up should be ensured.[17] But it is not clear whether Pakistan has thought through the implications of its doctrine and strategy beyond the generalised statements. According to one authority, "Pakistan's reliance on forward defence and counter-attack puts a premium on decisive action in the midst of the chaos of crisis."[18]

Since Pakistan's nuclear strategy is India-specific, the question that arises is: how will Pakistan's nuclear strategy stand up to India? Unfortunately, Pakistanis tend to dismiss the Indian doctrine of no-first-use and retaliation only as rhetoric; and this increases the risk of miscalculations and errors of judgement. But the central principle of the doctrine places the onus of escalation to the nuclear level on Pakistan. And at that level, the doctrine specifies "assured" retaliation. It is possible that Pakistan could misjudge and raise the stakes in a crisis. But given that India would have a sufficiently potent arsenal that would survive any Pakistani first strike and be capable of inflicting devastating punishment, the fundamental weakness of the Pakistani assumptions, leading to its nuclear doctrine and strategy, become clear. This is also the reason why there was so much nervousness among the Pakistani strategic community regarding the Indian debate and discussions on limited war within the nuclear overhang.

It is not intended to go into the details of Pakistan's nuclear programme and its present capabilities. Suffice it to say that from all accounts it has the wherewithal of a nuclear arsenal to achieve, at least theoretically, its nuclear strategy. The arsenal is estimated to include a possible stockpile of around 90-110 nuclear warheads. It possesses ballistic missiles supplied by China since the early 1990s. It has acquired ballistic missiles of ranges up to 1,500 km and claims developing the 2,500-km range Shaheen missiles. Its inventory of the F-16, J-10, Mirage III/V would serve to deliver nuclear bombs on most targets to around 400-km depth in western India.

Few, if any, political leaders (with the exception of Z. A. Bhutto), academics or security experts had articulated, leave alone debated the nuclear weapons or what Pakistan's nuclear strategy would be. Only the army (and its pro-army bureaucrats) asserted that Pakistan would follow a doctrine of first use, like, in fact, the two superpowers and their allies. It does not seem to have explored the doctrine and strategy of its all weather friend and mentor, China, which sought (successfully so far) to deter both superpowers at the same time (who between them had more than 65,000 nuclear warheads, pursuing the doctrine of first use) by a mere 400-odd nuclear warheads with only 20-odd Intercontinental Ballistic Missiles (ICBMs) with a "no-first-use" doctrine! Since then, China has modernised its arsenal but not really expanded it while Pakistan is reported to be furiously expanding its stock of nuclear warheads like the two superpowers did.

It is worth recalling that Bhutto was a strong advocate of Pakistan acquiring nuclear weapons and had started to urge Gen Ayub as early as 1958 to actively explore the nuclear weapons option. Ayub, however, had rejected Bhutto's advice on the grounds that if Pakistan needed a nuclear weapon, it could buy it "off-the-shelf."[19] Pakistan had started to face a disastrous defeat in the 1965 War triggered by Pakistan's Foreign Office under Bhutto, a covert war which the General Headquarters (GHQ) had strongly advised against launching into Kashmir on the grounds that India would react strongly. Gen M. Musa, the Army Commander-in-Chief, had endorsed this recommendation and the President, Field Marshal M. Ayub had asserted in writing that he would not let it happen. But it did happen and Pakistan rapidly faced a serious defeat. Ayub first sent Air Marshal Asghar Khan, the recently retired Commander-in-Chief of the Pakistan Air Force, to China to seek support where he was asked about how serious Pakistan was in fighting the war. When things were really getting out of hand, Ayub and

Bhutto made a secret visit to Beijing on September 19 night by air to request the Chinese to increase the pressure on India to save Pakistan. Beijing not only obliged, but also offered the bomb to Pakistan.[20]

But it was really after Pakistan's stunning military defeat in the 1971 War that the military utility of nuclear weapons became the central logic of Pakistan's nuclear policy.[21] Till 1971, the Pakistan military had always considered itself far superior to the Indian defence forces, both in fighting qualities and strategic and tactical operations. An ethno-professional arrogance appears to have dominated the belief system. However, the 1971 War demonstrated that not only were the Indian defence forces capable of fighting and defeating Pakistan but that this could be carried out in record time in the most difficult terrain of East Pakistan. At the same time, Pakistan was made to suffer a major defeat in the Western Sector. The end result after a fortnight was the emergence of Bangladesh as a sovereign state as a consequence of Pakistan's "tragedy of errors" and the highly, professional performance of the Indian armed forces. By the time fighting ended, over 90,000 prisoners of war were in the custody of the Indian forces, large tracts of West Pakistan were under the occupation of the Indian Army and Pakistan was split into two, dealing a deep blow to the two-nation theory that formed the *raison d'etre* of Pakistan to start with. The traditional friends of Pakistan, the United States and the People's Republic of China, had been of little help, and great disappointment.

Given this background, it is obvious that concerns in Pakistan would emerge regarding India's future policies toward Pakistan and the latter's ability to defend itself. The first one to emerge from that time was about the translation of India's inevitable conventional superiority into a decision on the battleground favourable to India. Expansion of India's military forces, triggered by the threat from the north and the Sino-Indian War of 1962, had resulted in a total force which would maintain a definitive quantitative edge over the Pakistani military in the future. This was indeed one of the factors that had propelled Pakistan (on Bhutto's prompting) to wage the war in 1965 since the Indian military would have become much more powerful once its expansion plans had fructified a few years later. The qualitative superiority was amply proved in the 1971 War.

Under these circumstances, it would be logical to assume that if India decided to retake Pakistan Occupied Kashmir (POK) militarily (which is part of the state of J&K which legally and constitutionally acceded to India, and, hence, POK remains a part of India), Pakistan might run into serious difficulties in trying to retain its hold there. After all, as Bhutto was to admit at Simla in July 1972, Pakistan had not been able to integrate that side of the state into the country even after a quarter century of occupation and had continued to maintain the façade of "Azad" (independent) Kashmir as an additional limb ruled by a Joint Secretary in the Pakistan government. The state of J&K had acceded to India in accordance with the provisions of transfer of power by the British. The UN resolution of August 13, 1948, had required Pakistan to vacate the territories occupied by it in the state. Given that Pakistan had no *locus standii* in occupying parts of J&K, the only means of ensuring continued control against the risk of Indian military action would be to possess nuclear weapons. The fact that Bhutto personally was convinced of the need from earlier days helped. But the central logic of the decision to acquire nuclear weapons was, no doubt, based on the perceived need to neutralise India's demonstrated conventional military superiority and protect its illegal possession of parts of J&K state. It would take time to build the bomb. In the meanwhile, reliance perforce had to be placed on diplomacy which in a few months took the shape of the Simla Agreement.[22]

Pakistan's drive (which soon became relentless), for nuclear weapons with the help of China and with the US facilitating it and supplying nuclear capable F-16s, had grown out of the enhanced sense of vulnerability after 1971 due to multiple factors:[23] the collapse of the ideology of the two-nation theory which constituted the very basis of Pakistan, the demonstrated superiority of Indian arms in a two-front war, the inability or unwillingness of its traditional allies to come to its assistance at a time when its very existence as a state was under question and risk, and the tenuous hold over the occupied portions of J&K. The combination of frustrations and the realisation that the very idea of Pakistan had come under question (by its own actions) coalesced into a burning desire for revenge. This gave hostility to India a new impetus where the realities of the power equation at that time did not provide the means to pursue its strategic objectives. Two crucial strands of Pakistan's strategy, without giving up its quest for occupying Kashmir, evolved after 1971:

- Nuclear weapons to neutralise India's conventional military superiority; and
- The use of irregular forces and *jihadi* terrorism for a direct assault on India (especially J&K), hopefully under the nuclear umbrella

It is in the above context that **Pakistan's first strike policy, especially at an early stage, can hardly be termed as a viable doctrine of deterrence though its nuclear weapons have served the role of blackmail/ compellence in tandem with *jihadi* terrorism. Pakistan using its nuclear weapons at an early stage of a conventional conflict would be suicidal for the country.**

**Nuclear Strategy Sans Credibility**

The then Army Chief of Pakistan had publicly declared the military doctrine in 1989 as that of "offensive-defence." When seen in the context of other indicators, this would appear to present a dual approach:

- At the tactical-operational level, it would seek to remain on the offensive through sub-conventional aggression, using the instrumentality of terror (for which Pakistan had acquired extensive means and expertise as a frontline state in support of the Mujahideen in Afghanistan and now in the US' war against terrorism) while the conventional military provides the defensive component. It may be recalled that when the offensive defence doctrine was articulated, it was also stated that the next war would be fought on Indian territory—as it turned out, through *jihadi* terrorists.
- At the strategic-operational level, the conventional military would provide the offensive component while the nuclear weapons provided the defensive element by (hopefully) deterring India's punitive action with conventional or nuclear capabilities. Most Pakistanis believe that their nuclear weapons capability has neutralised India's conventional superiority and, hence, hamstrung the Indian Army. This would be

expected to provide control over possible escalation. The expectation would be that the risk of escalation to the nuclear level would operate as a deterrent to any Indian counter-attack (of the type that was experienced in 1965 as well as 1971). The aggression in Kargil was obviously based on this philosophy.

These arguments may look good in briefings in operations rooms, sand model exercises or campaign documents. But with nuclear weapons around, the risk of failure is obvious. The basic factor that must be borne in mind always is that there is no historical or scientific template to assess the credibility of a nuclear strategy, especially if it is meant to be used first. While both superpowers relied on the doctrine of first use **during the Cold War, the credibility of nuclear deterrence was believed to rest on the second strike** and not really on the first strike which, in theory, was more effective for compellence. And the risks of first strike ranged from preemption to Launch On Warning (LOW), Launch Under Attack (LUA) and a constant risk of miscalculation. What will matter is the circumstances at that time, the rationality on both sides, India's actions and responses, and even the reaction of the international community that will have to be taken into account to assess the credibility of Pakistan's strategy and the theoretical doctrines. To begin with, therefore, if India does not impose a military situation where Pakistani leaders get worried about survival or even what they have termed as the "last resort," Pakistan's nuclear weapons and their credibility are nullified.

This, in fact, is the assumption of the Pakistani experts and military leaders: that Pakistan's nuclear weapons deter any Indian conventional military offensive. But does that really mean that India has lost its ability to wage a war against Pakistan, or to respond to a war, a regular or irregular one, that has been initiated by Pakistan? The answer is a categorical no. Hence, the second scenario is that Pakistan provokes a military response by India with a terrorist attack at a vital point (like the attack on the Parliament on December 13, 2001, or the November 26, 2008, terrorist commando attacks in Mumbai). Here comes the crucial aspect of a credible Indian strategy.

The Institute for Defence Studies and Analyses had been studying this issue since the late 1980s when Pakistan was believed to have acquired a credible nuclear weapon capability which it finally claimed in early 1987. The answer was clear: the Indian military response should not aim for a decisive military victory (which would inevitably lead to that "last resort" perception in the Pakistani GHQ). On the other hand, a number of options are available to apply force which hurts but does not necessarily destroy. In principle, therefore, our response will have to be a "limited war." Here it is necessary to point out that limited wars will produce limited military effects (which may or may not produce adequate political effects). One option to ensure a desirable level of punitive action has been to make multiple shallow penetrations across the border by the land forces instead of a single deep strike, a strategy that was in vogue in the 1980s before both sides acquired nuclear weapons, but has been made redundant by the existence of nuclear weapons.

It is important that the requisite domestic legitimacy for military action must precede such action. Hence, the principles and broad parameters of a limited war strategy would need to be widely disseminated and discussed without going into the details. At the same time, the political demands must be clear and achievable.[24] The limited war for punitive action would need to have the following characteristics:

- Aim to achieve politico-economic effects rather than only military effects which were central to wars for territory.
- Land forces would, of necessity, have to pursue a strategy of strategic defence and tactical offence, with limited penetration across the border. To be effective, multiple but limited penetrations across the border would have to be carried out. This would necessitate reorganisation of battle formations which, to some extent, has been done under the rubric of Cold Start.
- Employ the Indian Air Force for a strategic offensive for planned strikes preferably restricted to POK, while maintaining a strategically defensive posture outside J&K. Initially, small size targets may have to be selected to use the strikes as a signal for further action if the political objectives are not met
- Similarly, the Indian Navy would need to be deployed for punitive actions which would not lead to major naval actions.

One can go on along the above lines, but the principle of limited war in a nuclear environment would need to based on the fact that the adversary may escalate, but the initiative to further escalate must be retained in our hands. Here a "no-first-use" nuclear doctrine leading to **an assured counter-strike strategy** becomes the critical factor in providing the necessary credible deterrence against an adversary escalating beyond a certain level below the nuclear threshold. In short, military force can be applied below the nuclear level while a nuclear strategy would deter the early escalation to the nuclear levels by the adversary since the punitive action would take place (even if spread in time and space) well before the point of "last resort."

In short, Pakistan's first use doctrine and strategy do not comprise a credible strategic choice to avoid all types of use of force, especially limited war. And to come to that conclusion, one is not even going into all the extensive supporting infrastructure like space-based systems and high performance intelligence capabilities, rapid communications, etc. which all add up to an enormous amount of capital and running costs. Even then, the first use strategy has intrinsically far more risks of mistakes, miscalculations, etc. which could lead to unintended consequences threatening the very survival of all the people living in the region. This is why China adopted the local border war strategy by 1987 when both Pakistan and India had acquired nuclear capabilities and the US and Russia had long possessed them. And this is also the reason why China has formulated its military strategy to win local border wars through "command of the sea, command of the air and strategic forces."[25]

**Conclusion**

It is in the context of the above essay that this volume, the product of sustained, dedicated work in a complex area and objective analyses of facts as known at this time, by Dr. Shalini Chawla, serves the purpose of a landmark study to look into the different facets of a nuclear Pakistan. I feel privileged to have been asked to write an introduction to this valuable book which will no doubt find its niche among scholars, experts, practitioners of national defence and everyone interested in national security and defence. Nuclear weapons are different from conventional weapons if for no other reason than that they have been used only once, but are supposed to have kept the peace for over six decades. And the only way we can use them or defend against them is not on the basis of experience but by detailed studies and examination of potential scenarios to arrive at rational conclusions. I strongly recommend this book to everyone who wants to understand Pakistan's nuclear policy and/or is interested in peace and stability in the region.

**Notes**

1. M.A. Jinnah's address to the Constituent Assembly of Pakistan, August 11, 1947.
2. Stephen P. Cohen, *The Pakistan Army* (New Delhi: Himalayan Books, 1984), p.34.
3. Ibid., p.105.
4. Christina Lamb, *Waiting for Allah: Pakistan's Struggle for Democracy* (London: Hamish Hamilton, 1991), p.75.
5. Air Marshal A. Rashid Shaikh, PAF (Retd), "Security and Development: Hobson's Choice", *Defence Journal* (Karachi), Vol. XXI, May-June 1996, p.13.
6. Ross Masood Husain, "Threat Perception and Military Planning in Pakistan: The Impact of Technology, Doctrine and Arms Control" in Eric Arnett, ed., *Military Capacity and Risk of War: China, India, Pakistan and Iran* (Oxford: SIPRI/Oxford University Press, 1997), pp.130-147.
7. Cohen, n.2, p.145.
8. Brig S.K. Malik, *The Quranic Concept of War* (Lahore: WajidAlis Ltd, 1979), p.58.
9. This appears very similar to Al Qaeda's stated philosophy.
10. Malik, n.8, p.59.
11. Cohen, n.2, p.65.
12. Gen Mirza Aslam Beg, *The Nation*, December 13, 1993.
13. For a recent critique, see Altaf Gauhar, "Four Wars, One Assumption" *Nation*, September 5, 1999.
14. Agha Shahi, "Address," in Fasahat H. Syed, ed., *Nuclear Disarmament and Conventional Arms Control Including Light Weapons* (Rawalpindi: FRIENDS, 1997), pp.417-424.
15. "Securing Nuclear Peace," *The Nation* (Karachi), October 5, 1999.
16. M.B. Naqvi, *Defence Journal*, (Karachi), February 2000, p.15.
17. Gp Capt S.M. Hali, "The Advent of Air Power and Pakistan Air Force," *Defence Journal* (Karachi), February 2000, p.232.
18. Husain, n.6, p.131.

19. Cited in Samina Ahmed and David Cortright, "Going Nuclear: The Weaponisation Option," in Ahmed and Cortright, eds., *Pakistan and the Bomb: Public Opinion and Nuclear Options* (Notre Dame: University of Notre Dame Press, 1998), p.90.
20. Jung Chang and Jon Halliday, *Mao: The Unknown Story* (London: Jonathan Cape, 2005), p.518. It is generally believed that China had actually offered nuclear warheads which Bhutto was keen to accept. But Ayub reportedly set up a committee headed by Bhutto, with the Chairman of the Atomic Energy Commission and a Secretary of the government as members, that concluded that Pakistan could not afford to have the warheads and maintain them, technologically or financially.
21. Samina Ahmed, "Pakistan's Nuclear Weapon Programme," *International Security*, Vol 23, No. 4, Spring 1999, p.183.
22. It may be recalled that the Simla Agreement of July 2, 1972, was euphorically praised by Bhutto as the symbol and substance of ultimate peace in the subcontinent, and described by Abdul Sattar, then Foreign Minister of Pakistan as a "diktat."
23. Agha Shahi, Zulfiqar Ali Khan and Abdul Sattar, "Securing Nuclear Peace," *The News*, October 5, 1999.
24. This was the major flaw during Parakaram, where the political demands initially had little relevance to the mobilisation for military action. Similarly, all military actions and diplomacy have to be synergised for maximum effect.
25. *White Paper on National Defence 2004*, State Council of China, December 28, 2004. Here the strategic forces in the Chinese strategic calculus are likely to be used with manoeuvrable reentry vehicles carrying conventional warhead(s).

# 1. Pakistan's Nuclear Programme: The Build-up

Pakistan's nuclear history can be traced back to as early as the 1950s when the US President made the 'Atoms for Peace' proposal. The nuclear programme initially commenced with the objective of developing nuclear energy for civilian use, but Pakistan got serious about nuclear weapons after the 1971 War, which led to the dismemberment of Pakistan. After its third defeat against India, the Pakistani leadership was convinced that to neutralise India's conventional superiority, its only option was to acquire nuclear weapons. During, the initial years, from the early 1950s to 1971, Pakistan's official policy was to build up the nuclear programme for the development of nuclear energy. Starting from the 1970s under the leadership of Bhutto and subsequently in the coming decades, the development of the nuclear programme was shaped by increasing militarisation of Pakistan's state polices and its strategic alliances.

**The Beginning: 1950s**

Although Pakistan's formal nuclear programme can be traced to the 1970s following the defeat in the 1971 War; the efforts to go nuclear, with not necessarily an India-centric objective, but for the development of nuclear energy for peaceful purposes, began in the early 1950s. In December 1953, President Eisenhower made the 'Atoms for Peace' proposal and the proposal to establish the International Atomic Energy Agency (IAEA) was also launched. The Atoms for Peace received an extremely enthusiastic response in Pakistan. But no formal steps were taken due to the country's unstable political situation and also the poor knowledge and infrastructure in nuclear science and technology.

The US proposal gathered support from the international community and, subsequently, Pakistan's Foreign Minister also expressed support for the Eisenhower 'initiative'. In 1954, plans were developed to set up an atomic research body and in 1955, the Pakistani government announced its decision to establish an Atomic Energy Institute.[1] A committee known as the Atomic Energy Committee, headed by Dr Nazir Ahmad, was assigned the task of looking into, and acting upon, matters related to atomic energy. The mandate of the committee included: formulation of an atomic energy programme, estimation and identification of the requirements of its organisation, including the financial assistance in the form of loans from the foreign nations, survey of, and access to radioactive materials, and advice to the government on matters related to atomic energy research and its peaceful utilisation.[2] In 1956, the Pakistan Atomic Energy Commission (PAEC) was established under the chairmanship of Dr Nazir Ahmad on the recommendation of the committee.

Between 1954 and 1956, the US sponsored 'Atoms for Peace Exhibition' toured several Pakistani cities, propagating the use of atomic energy and provided Pakistan about 70,000 items of information about atomic energy', thus, creating a favourable public opinion about atomic energy.[3] The American initiative encouraged the domestic efforts within Pakistan, and the Government College of Lahore established the High Tension and Nuclear Research Laboratory to facilitate research for the post-graduate students in the Physics Department, to explore and develop the option of atomic energy.[4] The mandate of the PAEC aimed to plan and develop "peaceful uses of atomic energy, with special reference to survey, procurement and disposal of radioactive materials, planning and establishment of an Atomic Energy and Nuclear Research Institute installation of research and power reactors, negotiations with International Atomic Energy Agency bodies, selection and training of personnel, application of radioisotopes to agriculture, health, industry, etc."[5]

The commission's research work was started only in the late 1950s. The US cooperation was stepped up and the monitoring of radioactive fallout in select cities began, the instruction on nuclear science was upgraded and also Pakistan began to participate in the international conferences held under the United Nations and Commonwealth auspices.[6] The PAEC stepped up efforts in the direction of engaging and mobilising the academic community for the development of nuclear science in order to get human resources from the country for further nuclear establishments. The commission, as per its mandate, tried to encourage the development of atomic energy for agricultural, medical and industrial uses. Also, apart from exploring the presence of radioactive materials in Pakistan, it tried to acquire a power reactor to produce nuclear energy.[7]

To an extent, the commission succeeded in its aim of motivating the academic community and many Pakistanis were reported to have been trained in the foreign research institutions in the US and UK in the late 1950s and early 1960s. However, the commission was not very successful in mobilising the financial and political support within Islamabad to obtain a research reactor. The Pakistani government seemed to be lacking confidence in Nazir Ahmad's leadership. Also, the PAEC did not possess adequate trained personnel. In 1957, the PAEC completed the technical evaluation report of, and drafted a proposal for, the acquisition of the US CP5-type research reactor from the United States. The PAEC's proposal was vetoed by the Departments of Finance and Industry. The United States was not willing to supply Pakistan with a CP5-type of research reactor, which ran on heavy water, and offered a light water reactor instead.[8] The PAEC continued lobbying the Finance Ministry for resources in order to acquire the US reactor but its demands were rejected on fiscal grounds. The military elite in Pakistan did not show much interest in the 1950s for acquiring nuclear capacities.

**The 1960s**

In 1959, Dr I.H. Usmani, a physicist by training, succeeded Dr Nazir Ahmad and gave concrete shape to Pakistan's nuclear infrastructure by implementing the research reactor and power reactor plans. He maintained and further developed the orientation of Pakistan's nuclear programme towards peaceful purposes. Usmani worked towards the goals which his predecessor had failed to approach successfully. He had a more focussed approach for acquiring the research and power reactor. Also, he was able to get political and financial support from the government and get PAEC out of the bureaucratic hurdles. The commission under Usmani's chairmanship became the central authority to take decisions on nuclear issues. Another aspect upon which he stressed was training of the human resource for the development of nuclear science in the country. In fact, one of Usmani's most momentous achievements was the 'training programme' for which brilliant young Pakistanis were selected and sent abroad. Between, 1960-67, possibly around 600 personnel were selected out of whom 106 were reporteed to have returned with doctorate degrees.[9] Usmani's rational approach to nuclear energy for peaceful purposes is clearly projected in his statement:

> What the future of atomic energy would be, no one can guess. Its potential for good is immense; for evil, it is equally great. Its future, therefore, depends more on human endeavour. We stand on the cross-roads of destiny. A wrong turning may spell complete disaster and end the whole drama of life in smoke. Wisdom and prudence in the use of tools science has placed in our hands, may, on the other hand, help us carve out a future of lasting peace and prosperity. The choice had to be made, now or never.[10]

President Muhammad Ayub Khan fully supported the commission's activities for civilian use of nuclear technology. Also, at the international level, Pakistan's nuclear approach was welcomed and it received generous support from the US, UK, Canada and the IAEA. A grant of $3,50,000 came from Washington as assistance to prepare for Pakistan's first research reactor which the US agreed to supply after two years. In 1962, Pakistan acquired a 5 MW light water research reactor known as the Pakistan Atomic Research Reactor (PARR-1) which was located at the Pakistan Institute of Nuclear Science and Technology (PINSTECH) at Nilore, near Islamabad. The reactor began operating in 1965 and Pakistan produced the first batch of radioisotopes at PINSTECH.

In 1969, Pakistan also arrived at an agreement with Canada for the Karachi Nuclear Power Plant (KANUPP) power reactor. The inclination for French cooperation also came up in the 1960s. There were suggestions from the Pakistan Foreign Ministry to obtain a nuclear fuel reprocessing facility from France. Also, the French nuclear engineering firm Socie't'e G'en'erale pour les Techniques Nouvelles (SGN) offered to supply a 100-ton fuel reprocessing plant to PAEC. But both the offers were overruled by the Pakistani government.[11]

In the early 1960s, the Minister for Fuel, Power and Natural Resources and later Foreign Minister, Zulfiqar Ali Bhutto, was amongst the most enthusiastic advocates of the nuclear programme for Pakistan within Gen Ayub's Cabinet, but he was unable to convince Ayub and the government at that time to opt for a nuclear programme. Bhutto became much more expressive and vocal regarding his deep interest in the nuclear subject after he left President Ayub's government in 1966. The 1960s comprised a significant period in the history of Pakistan's nuclear programme, it witnessed the emergence of a sound nuclear policy in Pakistan and the leadership tried to generate political and diplomatic support for the nuclear programme.[12]

In India, the Chinese explosion in 1964 and its violation of the Comprehensive Test Ban Treaty (CTBT) sent shock waves, aggravating the insecurity which had loomed after the defeat in the Sino-India War, and caused great concern to the Indian policy-makers. The Indian leadership was always inclined for the use of atomic energy for peaceful puposes only. Lal Bhadur Shastri was, in fact, "horrifed", as Raj Chengappa terms it,[13] at the prospect of India developing nuclear weapons. Just before the Chinese blast, Shastri in a Non-Aligned Movement (NAM) conference at Cairo said:

> We in India stand committed to the use of atomic energy for only peaceful purposes and even though in a purely technical and scientific sense, we have the capability of developing nuclear weapons, our scientists and technicians are under firm orders not to make a single experiment, not to perfect a single device which is not needed for peaceful purposes.[14]

After the Chinese test, there was a tremendous uproar, which Bhabha anticipated, and on October 24, 1964 (just eight days after the Chinese test), Bhabha made his famous broadcast over All India Radio. Earlier, in his paper presented at the Third International Conference on Peaceful Uses of Atomic Energy at Geneva, he talked about a 10 kiloton device costing \$3,50,000 and a 2-megaton weapon costing \$6,00,000. In his broadcast on October 24, he stated that an inventory of 50 atomic bombs would cost under Rs 10 crore (or \$13 million) and a stockpile of 50 weapons of (2 megaton each) would cost around Rs 15 crore (or $25 million), which he said was very affordable for India.[15] He added, "The explosion of a nuclear device by China......is a signal that there is no time to be lost..."[16] Bhabha implied that the bomb programme was not as costly as had been projected. He also said that if the orders were given, India could make the bomb in just eighteen months.[17]

Pakistan's perceived threat from India was aggravated with these developments in India and propagated the notion in Pakistan that India was building the capability against Pakistan. Fear of India's conventional superiority in any case remained Pakistan's prime concern. Bhutto, in *Myth of Independence*, wrote:

> It is not possible for Pakistan, within the next few years, to develop a local industrial potential for equipping its armed forces with the more sophisticated weapons; nor can we depend entirely upon ingenious diplomatic initiatives.....Pakistan is a very dangerous affair for the aggressor, and we have this means to find an effective deterrence.[18]

He further added that Pakistan needed to give supreme importance to nuclear weapons and "Our problem, in its essence, is how to obtain such a weapon in time before the crisis begins."[19]

The Pakistani writings point to the developments in India in the 1960s and hold Indian initiatives responsible for the developments in Pakistan's nuclear programme. But the fact is that the Indian anxiety, to the extent it affected the senior leadership, was linked to the China factor. The Chinese nuclear test in 1964 became much more pertinent for India as the memories of the 1962 War were still very fresh. Bhabha's statement regarding India having the capability to make the bomb in 18 months is often highlighted in the context of a triggering factor in Pakistan's perceived threat from India. The statement was more an expression of India's capability in the nuclear field. In 1966, India spoke at the UN about its right to develop peaceful nuclear explosives under international supervision.

In the late 1960s, Pakistan experienced rising domestic unrest in its Eastern wing. The demand for provincial autonomy and resentment against

West Pakistan's discriminating policies increased in the late 1960s. On the other hand, after the 1965 India-Pakistan War, American sanctions against Pakistan led to the suspension of US aid and supply of military equipment. This was a major setback for the Pakistani defence modernisation which relied on the Western equipment. These factors in the country ruled by the army, no doubt, contributed in highlighting Pakistan's fears and tilting public opinion in favour of nuclear weapons.

**Policy on Disarmament and Arms Control**

Pakistan's policy on the issue of disarmament and arms control in the 1960s was in the formative stages. Pakistan signed the Partial Test Ban Treaty in 1963 but did not ratify it, pointing to the flaws in the treaty. Pakistan's stance was that the treaty prohibited nuclear tests in the atmosphere, outer space and underwater; but it failed to stop the nuclear states party to the treaty from further developing and refining their nuclear capabilities. Thus, Pakistan retained its nuclear option to test even though India signed the treaty and also ratified it.

In 1968, the nuclear Non-Proliferation Treaty (NPT) came into force but was unacceptable to both India and Pakistan. India rejected the treaty on the grounds that if provided the Nuclear Weapon States (NWS) a discriminatory right to develop, and control the use of, the nuclear weapon. The NWS would have the ultimate authority over nuclear arsenals, with no rights available to the Non-Nuclear Weapon States (NNWS) in this regard. Pakistan's multifacted stance for not signing the NPT has been briefly outlined by Ashok Kapur.[20]

First, Pakistan sought "unqualified adherence" to the treaty, especially by potential nuclear weapon states which, in the 1960s, meant Israel, India, South Africa, the Federal Republic of Germany and other important nations. Pakistan was never vocal in its stance that its lack of acceptance of the treaty was linked to the Indian position. But it made it very clear that it would not sign the NPT until the potential nuclear states join and adhere to it.

Second, Pakistan stated, "In the final analysis, the position of Pakistan with regard to signing the treaty will turn on considerations of its enlightened national interest and national security in the geo-political context of the region in which Pakistan is situated."[21]

The 'geo-political context' definitely included India but also it cannot be denied that Pakistan had other national security concerns which pointed towards the USSR. The Pakistan-Soviet relationship had not been positive due to the ongoing Cold War and Pakistan's anti-Soviet alliance with the USA resulted in the long desired flow of military support from the US.

Third, Pakistan's nuclear diplomacy desired security assurances for the NNWS. It demanded prohibition of first use by the nuclear weapon states against the non-nuclear states, and immediate assistance to be forthcoming, before the Security Council can act, for the non-nuclear weapon states which are victims of a nuclear attack. The security assistance asked for by Pakistan included all non-nuclear weapon states which renounced the manufacture or acquisition of nuclear weapons, irrespective of whether they had signed the NPT or not.[22]

Thus, Pakistan's nuclear diplomacy was not vocal about "India" in its rejection of the NPT, but it cannot be denied that India's arms control behaviour and Pakistan's apprehensions towards Indian intentions did shape Pakistan's position on the issue of the NPT.

**The Bhutto Era: The Islamic Bomb, 1972-77**

Bhutto made nuclear weapon capability a major public issue in spite of Ayub's antipathy to them. He argued the case for nuclear weapons from the time he became Foreign Minister in 1963. Gen Ayub resisted the idea and felt that Pakistan could not afford the bomb. He said, "What do we need a bomb for?........Pakistan is a poor country. We can't afford it,.....We should put our money into schools, may be hospitals, and industry."[23]

Bhutto was always in favour of Pakistan's acquisition of the nuclear capability. In his famous book *If I am Assassinated...* he wrote, "A country does not have to be merely wealthy to possess nuclear capability. If that were the only requirement, every OPEC country would have nuclear capability. The essential prerequisite is the infrastructure."[24]

The genesis of Pakistan's nuclear weapons capability dates back to 1965 when Bhutto made his famous statement that if "India builds the bomb", then Pakistan "will eat grass or leaves or even go hungry", but will build the bomb. Munir Khan, a Pakistani nuclear scientist, who was later appointed the Chairman of the Pakistan Atomic Energy Commission (1972-91), met Ayub on December 11, 1965, at the Rochester Hotel, London, but failed to convince him that Pakistan needed to take notice of India's nuclear programme and acquire nuclear capability. Reportedly, after the meeting, when Munir shared Ayub's views with Bhutto, he said, "Don't worry, our turn will come."[25]

He wrote later in *If I am Assassinated.....:*

> I have been actively associated with the nuclear programme of Pakistan from October 1958 to July 1977, a span of nineteen years. I was concerned directly with the subject as Foreign Minister, as Minister for Fuel, Power and Natural Resources and as Minister in Charge of Atomic Energy....... Assiduously and with granite determination, I put my entire vitality behind the task of acquiring nuclear capability for my country.[26]

Bhutto was the real architect of Pakistan's nuclear programme and after he left Ayub's Cabinet, he built his credibility on the basis of the nuclear build-up. It was during his tenure as the Minster In-charge for Energy and Natural Resources that PINSTECH and the Centre for Nuclear Studies (CNS) came up. The first stage of the PINSTECH reactor building and ancillary facilities was completed in 1965 and the second stage of PINSTECH, consisting of laboratories, workshop and auditorium, became operational in 1974, when Bhutto was in power.[27]

While Bhutto was out of power during 1966-71, he wrote extensively on the nuclear issues and its relevance to Pakistan's national security. His book, *Myth of Independence*, written in 1967 and published in 1969, portrayed his total conviction for building nuclear capability. He wrote: "Pakistan's security and territorial integrity are more important than economic development" and " the defence requirements of her sovereignty have to be met first."[28] Bhutto was determined to have the nuclear capability as a deterrent. His conviction for nuclear deterrence strengthened after the 1971 War, when Pakistan did not receive Chinese and US assistance and, thus, any international norms and treaties related to restricting Pakistan's nuclear weapon development were unacceptable to him. He asserted that Pakistan needed to acquire the nuclear bomb to match India's defence capability:

> All wars of our age have become total wars; all European strategy is based on the concept of total war; and it will have to be assumed that a war waged against Pakistan is capable of becoming a total war. It would be dangerous to plan for less and our plans should, therefore, include the nuclear deterrent. ............it is vital for Pakistan to give the greatest possible attention to nuclear technology, rather than allow herself to be deceived by an international treaty limiting this deterrent to the present nuclear powers. ................If Pakistan restricts or suspends her nuclear programme, it would not only enable India to blackmail Pakistan with her nuclear advantage, but would impose a crippling limitation on the development of Pakistan's science and technology.[29]

**1972: The Multan Meeting**

Systematic efforts for building the nuclear weapon commenced after the 1971 War. Bhutto called for a secret meeting of top Pakistani scientists in Multan on January 20, 1972, and, curiously, the surreptitious meeting to discuss Pakistan's bomb was held in a *shamiana.* Khalid Hasan, the former Press Secretary to Bhutto, recollected and narrated Bhutto's words in the meeting. He told the authors of *The Islamic Bomb,* that "Bhutto started slowly. He spoke of Pakistan's defeat and humiliation in the war with India, and vowed that he would vindicate the country's honour." He continued by saying that " he had always wanted Pakistan to take the nuclear road, but nobody had listened to him. Now fate had placed him in a position where he could make the decision, he had the people of Pakistan behind him, and he wanted to go ahead. Pakistan was going to have the bomb and the scientists sitting under the *shamiana* that day were going to make it for him."[30] Bhutto had asked Munir Ahmad Khan, who was with the International Atomic Energy Agency in Vienna, to prepare a report for Pakistan's nuclear programme. The report suggested slight progress on the nuclear programme to which most of the scientists present at the Multan meeting agreed.[31] The meeting brought out the frustrations and complaints of the scientists regarding the bureaucratic hurdles and lack of leadership retarding the growth of the nuclear programme. Bhutto, as usual, with his charismatic personality, was able to motivate the scientific community to get ready for building the bomb in three years instead of five, which the scientists had asked for. He received the enthusiastic response by the scientists, " We can do it.........you will have the bomb!"[32]

Soon after the Multan meeting, Bhutto started seeking external support to finance the nuclear weapon programme. He visited the Middle Eastern countries which included Iran, Saudi Arabia, Libya, Egypt, United Arab Emirates and Syria. Various reports in the past had suggested that the Arab world had been keen to develop the nuclear option. In 1971, Col Gaddafi of Libya sent his Prime Minister to China to buy the nuclear bomb. He also approached France, Canada, West Germany, and even India, seeking assistance in building nuclear weapons in response to his offer of oil at low prices. The quest for the nuclear weapons in the Arab world was understood as the Middle Eastern military balance had been predominantly in favour of Israel (suspected of testing a nuclear device in 1969) and the United States had made sure that the Israeli dominance in terms of conventional weapons would continue. This had led the Arab world, which had enough resources, to turn towards the nuclear option.[33] Thus, Bhutto's request for assistance received an enthusiastic response. He was looked upon as the champion of the Islamic cause. He was able to convince Saudi Arabia and Libya about the strong possibility of Pakistan being able to build a nuclear bomb as Pakistan's civilian nuclear programme had reached a level of erudition.[34] Though there is no direct evidence of the Libyan and Saudi support to Pakistan, it is believed that secret meetings were held in Paris to work out the modalities with representatives from Pakistan, Libya, Saudi Arabia and Iran. Bhutto did not have to wait too long for financial support from the Arab world. The price of oil went up extensively following the international oil crisis in 1973. The moment was appropriate for Bhutto's appeal for financial assistance to support the 'Islamic bomb'. In February 1974, he hosted an Islamic Summit in Lahore, which was attended by the leaders of 37 Muslim countries. The forum laid the ground for an alliance between the wealthy Muslim nations and poor Islamic countries, to fight the Zionists and their Western supporters.[35] According to some unidentified sources, as cited by Steve Weissman and Herbert Krosney (*The Islamic Bomb),* huge sums of money—close to a billion dollars—were raised for Pakistan. It is believed that Gaddafi and Bhutto came to a formal agreement in February 1974 during the Islamic Summit hosted by Bhutto in Lahore, and Libya agreed to finance the entire Pakistani nuclear programme in return for the first bomb that was to be made. According to Mohammad Beg, a former Pakistani diplomat and journalist:

> Mr. Bhutto told me that during the Islamic Summit in Lahore, he had several discussions with Colonel Gaddafi about the manufacture of an atomic bomb by Pakistan…..[Gaddafi] promised Mr. Bhutto that he would provide however much money was needed. And, in return, he asked Mr. Bhutto if Libya could have the first bomb.[36]

Reportedly, in a few months, two couriers from Libya reached Pakistan, each carrying $100 million in cash. The King of Saudi Arabia also offered financial support for the bomb and, in return, expected Pakistan to provide a security umbrella for the kingdom.[37]

Libya, in return, desired complete access to nuclear technology and also wanted the Pakistani scientists to train Libyan technicians in the operation of hot cells, neutronics and the handling of nuclear waste.[38]

### Failure to Obtain the French Reprocessing Plant and Emergence of A.Q. Khan

The Canadian plant, KANUPP, established in Karachi, was on stream and not only produced electricity but also plutonium, which obviously had potential military uses. Pakistan had to rely on Canada for expertise, fuel and spare parts, and in return KANUPP was under IAEA safeguards. On May 18, 1974, India carried out a PNE (Peaceful Nuclear Explosion). International proliferation concerns increased and Canada stopped the supply of nuclear fuel to the plant. Bhutto had requested France in 1972 for the procurement of a reprocessing plant. The French government responded positively, and in 1976, Pakistan signed an agreement with France to acquire a plutonium reprocessing plant, to be installed at Chashma under an international safeguards agreement signed by Pakistan, France and the IAEA. Though the plant was supposed to be under IAEA safeguards, the Americans were deeply concerned about the fact that Pakistan could use the reprocessing facility for developing nuclear weapons. US Secretary of State Henry Kissinger tried to convince Bhutto to stop the building of the reprocessing plant. Bhutto's refusal to cancel the French deal and insistence that the French plant would be used for peaceful purposes, was taken as an insult to the American intelligence by Kissinger. Bhutto said:

> Dr Henry Kissinger, the Secretary of State for the United States, has a brilliant mind. He told me that I should not insult the intelligence of the United States by saying that Pakistan needed the reprocessing plant for her energy needs. In reply, I told him that I will not insult the intelligence of the United States by discussing the energy needs of Pakistan, but by the same token, he should not discuss the plant at all.[39]

Failing to push Bhutto, Kissinger made a trip to Paris and was successful in persuading the French not to go ahead with the plant. Subsequently, in 1977, President Giscard D'Estaing agreed to cancel the deal.[40] The failure of the reprocessing plant deal was a major blow to Pakistan's ambition and Bhutto's plans to go ahead with the bomb looked difficult to materialise; thus, *enriched uranium* as an alternative for plutonium was seriously considered by the Pakistanis.

It was at this point that Dr A.Q. Khan, the founder of Pakistan's nuclear programme, commenced his contribution to the programme and brought in the gift of centrifuge-based enrichment technology. Dr A.Q. Khan was projected in the media throughout Europe as a man of immense mystery—"an Islamic blend of James Bond and Dr. No, using the magic of the East to steal the secrets of the West." Khan was proclaimed as the "spy who stole the bomb for Islam"[41] Dr Khan who had studied metallurgical engineering, was a recent graduate from a Belgium University. In 1972, he joined Fysisch Dynamisch Onderzoek (FDO), a subsidiary of Verenidge Machine Fabrieken, a company which worked closely with Ultra Centrifuge

Nederland (UCN), the Dutch member of the Urenco uranium enrichment consortium.[42] Dr Khan served as sub-contractor and consultant at the ultracentrifuge process which called for the finest precision. This unique opportunity strengthened his knowledge of, and experience in, the use of ultracentrifuges in the enrichment of uranium.[43] Though Khan was officially engaged in limited metallurgical research for the ultracentrifuge project, due to the relaxed security rules and also the confidence he gained from the Dutch authorities, he visited Almelo several times and also had access to Urenco information. He had complete access to Almelo's design plans and also inside knowledge of the firms supplying components for the ultracentrifuges.[44] The most important knowledge that he gained was about the European manufacturers in the nuclear sector. Khan acquired the crucial knowledge of centrifuge operations while working at FDO. He had picked up extensive information about the requirements for building a uranium enrichment plant. He had managed to smuggle the classified information and also the centrifuge designs from Almelo and, thus, paved a uranium enrichment path for Pakistan's nuclear weapons. This was something none of the developed nations had expected. Few Western nations had mastered and built their own enrichment plants as the technology was expensive and, more importantly, the construction of these plants proceeded secretly.

In 1974, Khan approached Bhutto with the intention of returning to Pakistan to use his experience in uranium enrichment to make a bomb. The Indian explosion in 1974 enthused Khan's patriotism and he was keen to get back to Pakistan and offer his services. In August 1974, he wrote his first letter to Bhutto, stating that he was an expert in uranium enrichment, working on advanced designs for the European Consortium. His first letter was ignored by Bhutto, and Khan wrote his second letter on September 17, 1974, which caught the attention of Bhutto, who was then desperate due to the sluggish movement on the plutonium project.[45]

Khan's return to Pakistan in the fall of 1975 was a great relief to Bhutto who could now be hopeful about the nuclear weapon despite the Canadian and French disappointments. Khan worked within the PAEC structure for some time and eventually could convince Bhutto about having his own independent establishment for producing weapon grade uranium and, consequently, the *Kahuta Laboratory* was born. Bhutto brought in the military's Special Works Organisation to oversee the construction at Kahuta and to help with the purchases. Khan was quite familiar with all the companies and his networking was helpful in getting access to the equipment. The major purchases were done from Germany, France and Great Britain for building the nuclear infrastructure to make weapons. And, thus, Khan was able to handle one of the most ambitious engineering projects, building a little Almelo in Pakistan.[46]

By 1975, the efforts of the Western government to control the illegal proliferation of nuclear weapon related equipment had started. An international group, the *London Suppliers' Group*, was formed which imposed an embargo on the supply of nuclear material and technology to any nation (like Pakistan) which was a non-signatory to the NPT. The US was also constantly warning Pakistan that it would cut off all American assistance to Pakistan if it continued to go ahead with its dealing in nuclear equipment. The US was aware that Bhutto was travelling to the Middle East in order to gain financial support for his nuclear projects.[47] In 1976, the United States enacted its non-proliferation law, the Symingyton Amendment to the Foreign Assistance Act of 1961. The law prohibited US economic and military assistance to any nation importing enrichment or reprocessing technology unless the nation was following the IAEA-NPT safeguards. Pakistan now faced the suspension of US assistance on account of the fact that it had imported the reprocessing technology from France.[48]

During 1977 and 1978, there were significant domestic developments within Pakistan. Although Bhutto was not unpopular, the nation remained restless during his tenure. His economic polices failed to produce positive results and his autocratic tendencies had seriously started to interrupt the governance of Pakistan. Bhutto reduced the power of the civil services and didn't allow the growth of the political institutions. The military was, anyway, unhappy with its undermined role and fully aware of its capacity to take over control of Pakistan. With political institutions being weak, the military was successful in crafting the departure of the Pakistan People's Party (PPP). In 1977, Zia-ul-Haq, Army Chief of Staff, chosen by Bhutto over ten more senior Generals in 1976, led the military coup and declared

Martial Law, announcing the return of military rule in Pakistan. And, the nuclear programme was moved from civilian to military control.

Bhutto's era is important in the history of Pakistan's nuclear development mainly due to two factors, Firstly, Bhutto provided a new dimension to Pakistan's nuclear endeavours. He successfully inspired the scientific community, and the nuclear programme became aggressive and India-centric. The factor of India-centricity evoked favourable public opinion for the nuclear capability build-up. Secondly, Bhutto managed to establish crucial links with the Arab world and was able to swing the much needed funding for the nuclear weapons. His efforts added to Pakistan's image as the "leader of the Muslim world".

**Gen Zia: Posture of Nuclear Ambiguity**

Pakistan's nuclear weapon programme was further built up on the existing infrastructure with similar passion under the leadership of Gen Zia-ul-Haq, As Zia-ul-Haq stated in July 1978, "China, India, the USSR, and Israel in the Middle East possess the atomic arm. No Muslim country has any. If Pakistanis had such a weapon, it would reinforce the power of the Muslim world."[49]

Unlike his predecessor Bhutto, who adopted a confrontationist attitude, Zia was relatively subdued in his approach and did not formally declare his intention to acquire the nuclear bomb. Gen Zia created a posture of nuclear ambiguity for Pakistan. He mounted a peace offensive to join the NPT on the condition of India signing it.[50] The military regime in Pakistan tried to link the Indian threat with acquisition of nuclear capability to gain domestic legitimacy for the bomb. The scientific establishment, functioning on the directives of the military, justified the threat perception and the military regime obviously fully supported them in every possible form, including generous funding.

The efforts for building the bomb continued and a vast and audacious purchasing campaign began. Reports suggest that between 1977 and 1980, Pakistan imported the entire plant for converting uranium powder into uranium hexafluoride from West Germany for the Kahuta enrichment facility. The German plant, like the Kahuta facility, remained outside the IAEA safeguards. The UK supplied Pakistan high frequency inverters to control centrifuge speeds.[51] Due to Khan's efforts, and the laxity in the Western export control measures at that time, Pakistan's nuclear programme made good progress in developing U-235 capability. According to Khan (in an interview in 1998), the first enrichment was done at Kahuta on April 4, 1978. The plant was made operational in 1979, and by 1981, was producing substantial quantities of enriched uranium.[52]

In the 1970s, the United States and Great Britain made serious attempts to prevent Pakistan from acquiring nuclear weapons technology. The US and British authorities became aware of Pakistani efforts to purchase the technology to produce fissile material for nuclear weapons—plutonium and highly enriched uranium. [53] Reportedly:

> In November 1978, the United Kingdom and the United States sent complementary demarches to other members of the Nuclear Suppliers Group (NSG) in efforts to 'delay' the Pakistani nuclear program by denying it access to sensitive technology and equipment. The US demarche was the first of about 300 sent over the next three years. In its first such demarche, Washington was trying to halt a secret Pakistani effort to continue a plutonium reprocessing facility at Chashma, which the French had begun but backed out of, partly in response to US encouragement.[54]

With this growing proliferation, the United States tried to persuade Zia to stop the activities at Kahuta. Zia refused to accept IAEA safeguards on Pakistan's nuclear establishments and also declared that Pakistan's nuclear programme was for peaceful purposes, which has been the official stance of Pakistan. In 1979, the Symington Amendment was imposed and the US economic assistance to Pakistan was terminated. But very soon, the United States' non-proliferation measures were undermined by its own interests.

The Soviet invasion of Afghanistan in December 1979 brought about renewed US-Pak engagement and Pakistan became the frontline state for the United States in the fight against Communism. Pakistan felt deeply threatened with a destabilised Afghanistan and the presence of Soviet military forces on its western borders. Along with concerns about the eastern border, the turmoil in Afghanistan made Pakistan's western borders also insecure. The status of being a US ally provided Islamabad with an excellent opportunity to demand substantive aid and the latest military equipment. Some issues were unanimously agreed between the military leadership and Washington: firstly, Pakistan would not compromise on its nuclear programme; secondly, US interference in Pakistan's internal matters was not acceptable and it was solely Pakistan's prerogative to solve its internal affairs.[55] Washington lifted the sanctions imposed on Pakistan and offered substantive military aid and equipment to Pakistan. Pakistan also received unlimited funds from Saudi Arabia and the Gulf countries to finance the US weaponry. This provided enough space to the military leadership to expand the nuclear programme.

Tom Weiner stated in 1998:

> The United States provided Pakistani nuclear scientists with technical training from the 1950s into the 1970s. And it turned a blind eye to the nuclear weapons program in the 1980s, because Pakistan was providing the crucial link in the Central Intelligence Agency's effort to smuggle billions of dollars of weapons to the Afghan guerrillas attempting to drive out the Soviet invaders.[56]

The renewed cooperation between the US and Pakistan which allowed the flow of Western equipment into Pakistan and invoked exemption to the Symington Amendment raised the level of discomfort in India. In 1982, in an interview, Indira Gandhi overtly displayed her unhappiness with Washington with regards to the developments in Pakistan's nuclear programme:

> The knowledge we have gained purely from the Western sources, American and otherwise, is that they (Pakistan) are going in for a bomb. Now, America has accepted that situation, so far as we can make out.[57]

**China's Role**

China's role during this period was extremely critical in the development of Pakistan's nuclear programme. According to Gary Milhollin, a leading expert on proliferation:

> China, a staunch ally of Pakistan, provided blueprints for the bomb, as well as highly enriched uranium, tritium, scientists and key components for a nuclear weapons production complex, among other crucial tools. Without China's help, Pakistan's bomb would not exist…[58]

China emerged as a trusted ally for Pakistan in the 1960s, and the Sino-Pak defence cooperation can be traced back to 1963 when the two countries signed the Shaksgam Valley Agreement. Chinese conventional arms began to flow into Pakistan after the 1965 India-Pakistan War when the US arms embargo suspended the Western equipment into Pakistan. Since then, China has played a critical role in Pakistan's foreign policy and its drive to possess nuclear weapons, and both countries have complemented each other's strategic requirements. Reportedly, substantive nuclear cooperation between the two started in 1976 with a meeting between Bhutto and the Chinese leadership, and the understanding was that while China would ultimately provide Islamabad with the design of a nuclear weapon and 50 kg of highly enriched uranium, nuclear sharing was reciprocal.[59]

In 1986, the Comprehensive Nuclear Cooperation Agreement was signed between Pakistan and China. The agreement was concluded for an initial period of 30 years and thereafter automatic extension of five years, unless the agreement is terminated by either side. Article II (f&h) of the agreement lays out that the two parties shall cooperate in the "design, construction and operation of nuclear research and power reactors and associated facilities, and other fields as may be mutually agreed upon."[60] According to the agreement , "China would construct four nuclear plants in Pakistan, namely: Chashma 1, 2, 3 and 4 by 2011."[61]

In the 1980s, various reports suggested Chinese assistance to the Pakistani enrichment programme, although Beijing consistently denied the reports. China involved the Pakistani scientists in a nuclear test at its Lop Nur test site in 1983. Reports suggest that Pakistan's Foreign Minister Yaqub Khan was also reportedly present at the site. The US was increasingly suspicious, even sure, about the Chinese assistance to Pakistan's nuclear programme. In 1983, a State Department Report suggested: "We believe that Pakistanis have experienced difficulty in making their centrifuge machines work and that the Pakistanis have not yet produced any significant quantities of enriched uranium. Because of these operational problems, the Pakistanis in the recent past sought help from the Chinese."

The reports further stated:

> We have concluded that China has provided assistance to Pakistan's program to develop a nuclear weapons capability. Over the past several years, China and Pakistan have maintained contacts in the nuclear field. For some time, China's involvement was limited to operational aspects of the KANUPP power reactor at Karachi. We now believe cooperation has taken place in the area of fissile material production and possibly also nuclear device design.[62]

Information from various sources indicates that Pakistan had negotiated the deal for the M-11 during Zia's regime. According to US officials, Pakistan agreed to pay China $15 million as partial payment on its 1988 contract with the state-owned China Precision Machinery Import-Export Corporation (CPMIEC) for an undetermined number of M-11 missiles, launchers, and support equipment.[63] In the same period, the National Development Complex (NDC), a subsidiary of the Pakistan Atomic Energy Commission (PAEC), reportedly acquired complete though unassembled M-11s and possibly an undisclosed number of M-9 Short Range Ballistic Missiles (SRBMs) from Beijing.[64] Although Pakistani leaders have made conflicting statements regarding the receipt of the missiles, various reports provide enough evidence that Pakistan did receive the M-9 and M-11 missiles from China. The M-9 SRBMs, which are capable of carrying both nuclear and conventional warheads, were delivered to Pakistan in early 1991.

The decade of the 1980s proved to be a blessing for Pakistan and the military was able to fulfill its long desired dream of acquiring high technology conventional military equipment from the West. Pakistan continued its nuclear related procurements and uranium enrichment during this period. Intelligence reports suggested that Pakistan had acquired uranium concentrate from Libya, vessels and tanks of special metal from

Italy, and a complete design of a 25 kiloton nuclear bomb from China.[65] In 1978, A.Q. Khan had enriched uranium to 3 per cent and was moving towards 90 per cent (the amount required for the weapon) which was not an easy task. In the early 1980s, a small-scale pilot plant was set up near Multan in Dera Ghazi Khan, with the assistance of a German company—CES Kalthof, GmbH. This reportedly, gave Pakistan the capability to produce various fluorine components, including uranium hexafluoride.[66] By 1984, Zia himself announced that Pakistan has acquired the capability of producing small quantities of low enriched uranium. In 1981, 1,000 centrifuges were being assembled and they could produce uranium for one nuclear weapon every two years. In the following couple of years, Pakistan had acquired up to 10,000 centrifuges which could produce 150 kg of enriched uranium per year and which would amount to seven nuclear bombs every year.[67] According to a former senior Pakistani official, Pakistan had acquired complete nuclear capability by 1983. Zahid Hussain in his book claims that Dr Khan openly announced in an interview to him that his organisation had conducted several cold tests in 1983 and 1984. Dr Khan apparently also claimed that Gen Zia was aware of the fact that Pakistan was ready for a nuclear test at a week's notice.[68] By 1986, Pakistan is thought to have produced enough fissile material for the weapon and by 1987, it is believed that Pakistan could carry out a full nuclear explosion. A.Q. Khan in an interview to Kuldip Nayar said that Pakistan possessed the bomb.[69] The interview was deliberately arranged as a signal to the Indian leadership about the consequences of militarily attacking Pakistan. Various reports also suggest that Pakistan conducted cold tests of a nuclear implosion device at Chagai in 1986.

Zia adopted a policy of development rather than declaration. Even though he had full intention of going for the nuclear bomb, he never voiced it. He knew that doing this would deprive Pakistan of the international support that it had acquired. He got a well established infrastructure in place, due to his predecessor's efforts, and he worked on further enhancing it by acquiring the funds and the nuclear technology and materials. The accelerated nuclear programme during Zia's time was to match the Indian capability but Zia never confronted it publicly.

**The 1990s: Nuclear Pakistan**

In 1988, after the death of Gen Zia in a mysterious plane crash, the military regime reverted to a democratic set-up in Pakistan in the wake of growing domestic resentment against military rule. The elections in 1988 led to the installation of an elected government. But the command and control of defence and security matters, including the nuclear issues, remained with the military. This can be attributed to the strong military set-up in the country and also the weakness on the part of the democratically elected leaders to assert their power over these matters.

Pakistan's nuclear related procurement continued despite changes in the leadership at the national level, although Pakistan kept convincing the US government that it had no intention of building a bomb and its nuclear programme was for peaceful purposes. Pakistan continued to maintain its position on the NPT and refused to sign it unless India became a signatory. In the wake of Pakistan's growing nuclear programme, in 1990, the Bush Administration imposed sanctions under the Pressler Amendment of the US Foreign Assistance Act, suspending all US military and economic assistance to Pakistan. The second US aid package worth $4.02 billion, with half of it as the military component, was for a period of six years, commencing in 1987, was suspended in 1990. The Pressler Amendment disappointed the Pakistanis and dried up the American source of supply to Pakistan. But it had little negative impact on the nuclear programme.

Gen Aslam Beg, in his writings, has talked about Pakistan's anger at the United States after the sanctions were imposed. According to him, Washington had placed Richard Barlow, an American in Islamabad, to monitor Pakistan's nuclear programme. Richard Barlow, asserted in his report that Pakistan had acquired the nuclear capability in 1987 itself. The report by Richard Barlow was widely circulated but did not receive much attention in the United States government, presumably because at that time the US needed Pakistan. Thus, from 1987-89, the US President continued to certify that Pakistan did not possess nuclear weapons. But, after the Soviets withdrew from Afghanistan in 1989, Washington became uncomfortable with Pakistan's possession of the nuclear capability and Mr Richard Barlow's report was used as testimony and Islamabad was accused of crossing the "Red Light". Subsequently, sanctions were imposed, and military and economic aid was suspended, inflicting a significant blow to Pakistan's economy and weapon modernisation. Pakistan faced tremendous pressure from the Bush government to suspend its nuclear activities, abandon the nuclear mission and get back to its original nuclear programme for peaceful purposes.[70] Gen Mirza Aslam Beg stated, " Pakistan had developed this capability in 1987, but the 'slings' and 'arrows' against it were triggered in 1990 for the ostensible reason that the Soviet forces had left the Afghan soil. The US security scenario had metamorphosed and Pakistan was the chosen target for the strangulation of its 'nuclear neck'."[71]

Pakistan at that time had embarked on all the nuclear stages simultaneously:

- The enrichment of the uranium up to 95 percent and above.
- Preparation of the device by integrating all its elements.
- The laboratory test of the different elements of the device.
- The final test of the integrated system. This can be accomplished either by detonation, which is termed a 'Hot Test' or through a 'Cold Test' without it.
- Developing the delivery system through special airplane or missile technology.[72]

Pakistan spent $250 million between 1979-92 for its nuclear programme, which according to its former Army Chief, Gen Aslam Beg, provided it with a credible "nuclear deterrent".[73] The US asked for the programme to be frozen. The US not only sought freezing but also rolling back of the programme and the demolishing and dismantling of all the nuclear facilities. This was obviously tough for Islamabad to accept and led to hardening of the Pakistani attitude. The United States was loud and clear about the apprehensions the West would have about the *Islamic bomb.* By 1992, Pakistan declared officially that it possessed the cores of nuclear weapons. On the other hand, China continued to provide the aid and delivery capabilities for Pakistan's nuclear programme despite Beijing's assurances to Washington. There is general agreement that China provided Pakistan with the blueprints to the uranium implosion device that China detonated on October 27, 1966 (CHIC-4 test/design).[74] Reports also suggest that on May 26, 1990, China tested a Pakistani derivative of the CHIC-4 at its Lop Nur test site, with a yield in the 10-12 kiloten (kt) range.[75]

The United States imposed sanctions on both China and Pakistan in 1991 for China's transfer of M-11 missile-related equipment. The sanctions were lifted in 1992 before they were reimposed on August 25, 1993, against certain Chinese and Pakistani entities that were involved in the transfer of M-11missile related equipment. The US sanctions against China were lifted in 1994. The US sanctions had failed to deter Chinese assistance to Pakistan.[76] The Central Investigation Agency (CIA) had notified the State Department that China had sold 5,000 ring magnets to the A.Q Khan laboratory in Kahuta. The Chashma nuclear power plant project which was previously initiated and cancelled by France in the 1970s, was resumed in the early 1990s. The nuclear plant was under IAEA safeguards, with the China Nuclear Energy Industry Corporation (CNEIC) as the foreign supplier, and the reactor was based on China's first indigenous reactor, Qinshan-1.[77] China did pledge in 1991 and 1994 that it would comply with the Missile Technology Control Regime (MTCR) and would abstain from missile technology transfer. But, in December 1992, reports suggested that China has apparently transferred 34 complete M-11 systems to Pakistan. The reports were obviously denied by China.[78]

Benazir Bhutto, who came into power twice in the 1990s, while not involved in the intricacies of the nuclear programme, fully supported it, in cooperation with North Korea and China. Pakistan was indeed keen to acquire the ballistic missile capability of striking strategic Indian targets. Benazir visited Pyongyang after having talks with North Korean President Kim Il Sung. She was accompanied by Foreign Minister Sardar Asef Ahmad Ali and Minister of Defence Aftab Shaban Mirani. According to Pakistani officials, the Pakistani delegation went with plans for North Korea's No-dong missile.[79] While Pakistan's clandestine missile transactions with China have been closely monitored by the US, the same does not stand true for its clandestine missile technology transfer from North Korea which has enjoyed greater leverage. The cooperation with North Korea amounted to Pakistani efforts to manufacture the No-dong missile under the name Ghauri (Hatf-5)[80] Cooperation with North Korea was amplified and most of the missiles and supporting equipment were delivered to Khan Research Laboratories (KRL) in the spring of 1996 by the Changgwang Sinyong Corporation (a.k.a., North Korea Mining Development Trading Corporation/Bureau). On April 24, 1998, the US State Department imposed sanctions for the second time against both KRL and Changgwang Sinyong Corporation.[81] Ironically, most of the cases of transfer of nuclear technology occurred during the 1990s, the decade of US sanctions on Pakistan.

On May 28 (and May 30), Pakistan conducted nuclear tests in the Chagai mountains in Baluchistan province. The Pakistani tests followed the Indian nuclear tests on May 11, 1998, which caught Pakistan by surprise. Since Pakistan already had the complete capability of the nuclear bomb (by 1987), it was not difficult for it to have the tests at short notice.

The tests provoked strong international condemnation. Pakistan claimed that its actions were reactive to the Indian tests and necessary for the survival of Pakistan. But the fact is that Pakistan has been pursuing the nuclear path with deep dedication from the early 1970s, primarily to compensate for asymmetry in conventional capability and to use nuclear weapon capability as an umbrella to conduct its strategy of covert war against India.

In 1998, when Pakistan conducted the nuclear test, it was already subject to limited sanctions imposed under the Pressler Amendment, which banned military and humanitarian aid to Pakistan. After the nuclear tests, President Clinton announced additional sanctions as required by the Arms Export Control Act and Export-Import Bank Act. The sanctions resulted in the termination of US foreign assistance other than humanitarian or food assistance, and termination of US government sales of defence articles and services.[82]

**Current Status of Nuclear Power**

After the overt nuclear tests, Pakistan has continued to expand its nuclear arsenal, both the number of warheads as well as delivery systems, including ballistic missiles, cruise missiles and nuclear capable aircraft.[83]

According to the *World Association of Nuclear Operators*, the current status of nuclear power in Pakistan is as follows[84]:

- Pakistan first nuclear reactor, KANUPP, supplied by Canada, is a small 137 Mwe (125 Mwe net), Pressurised Heavy Water Reactor (PHWR). The reactor was started in 1971 near Karachi and is operated at reduced power.
- The second reactor, Chashma-1 (also known as CHASNUPP-1) in Punjab, supplied by China under safeguards, was started in 2000. Chashma-1 is a 325 Mwe (300 Mwe net) 2-loop Pressurised Water Reactor (PWR) (Table 1.1). The main part of the reactor was designed by the Shanghai Nuclear Engineering Research and Design Institute (SNERDI) based on Qinshan-1. Construction for Chashma-2 commenced in 2005 with the Chinese technical and financial assistance. The enriched fuel for the PWRs is imported from China.
- In 2008, Pakistan announced plans for the construction of two new reactors, Chashma-3 and 4, with Chinese assistance. China has agreed to provide 82 percent of the total cost at an extremely low interest rate. The construction agreement between China and Pakistan was signed in 2010 for the two 340 Mwe (315 Mwe net) units that are scheduled to be completed in eight years (Table 1.2).
- Reportedly, in 2010, PAEC has signed a construction agreement with China National Nuclear Corporation (CNNC) for the fifth unit at Chashma.
- As far as the fuel cycle is concerned, a uranium centrifuge plant at Kahuta has been under operation since 1984. The Kahuta project was launched in 1972 after the India-Pakistan War in 1971. Reportedly, there is also a new plant at Gadwal.
- In 2006, PAEC announced that it was preparing to set up a new civil conversion, enrichment and fabrication plants—the Pakistan Nuclear Power Fuel Complex (NPFC) for PWRs reactors.
- PARR-1 is a 10 MW pool-type research reactor (of 1965 vintage), supplied by the US under the Atoms for Peace Programme. In 1991, the reactor was converted to use low enriched uranium fuel. PARR-2, operational since 1974, is based on the Chinese design and is using high enriched fuel. Both PARR-1 and 2 are located at the PINSTECH Laboratory, Nilore, near Islamabad.

- In 1998, a larger multipurpose reactor started operating near Khushab. The reactor is reportedly producing weapons grade plutonium. Another large heavy water reactor has been under construction at Khushab since 2002 and was supposed to become operational in 2009. In 2006 and 2011 (supposedly), the construction of the third and fourth reactors commenced. These would certainly add to substantive plutonium production in Pakistan which would compensate for the laxity in uranium resources.

**Table 1.1: Operating Nuclear Power Reactors**

| Reactor | Type | MWe Net | Construction Start | Commercial Operation | Planned Close |
|---|---|---|---|---|---|
| KANUPP | PHWR | 125 | 1966 | 12/72 | |
| Chashma-1 | PWR | 300 | 1993 | June 2000 | 2040 |
| Chashma-2 | PWR | 300 | 2005 | May 2011 | 2051 |
| Total (3) | | | | | |

*Source:* "Nuclear Power in Pakistan" at http://www.world-nuclear.org/info/inf108.html

**Table 1.2: Nuclear Power Reactors Under Construction, Planned and Proposed**

| Reactor | Type | MWe Gross | Construction Start | Commercial Operation |
|---|---|---|---|---|
| Chashma-3 | PWR | 340 | May 2011 | 2016 |
| Chashma-4 | PWR | 340 | 2012 | 2017 |
| Chashma-5 | PWR | 1000? | | |
| Total 3 | | | | |

*Source:* "Nuclear Power in Pakistan" at http://www.world-nuclear.org/info/inf108.html

Islambad's existing nuclear infrastructure, with the construction of additional reactors and reprocessing facilities, indicates its plans to expand the number of warheads. It is also developing new delivery systems. With the current rate of production, it has the fastest-growing stockpile in the world today; it is estimated that Pakistan has a nuclear weapons stockpile of 90-110 nuclear warheads, which is a steep increase as compared to the 70-90 warheads it possessed in 2009.[85]

## Notes

1. Ashok Kapur, *Pakistan's Nuclear Development* (London: Croom Helm, 1987) p. 35.
2. Ibid.
3. Ibid., p. 36.
4. Ibid., p. 36.
5. Dr Nazir Ahmad, "The Atomic Energy Commission", *Pakistan Quarterly*, Vol. 7, No. 3, Autumn 1957, p. 14, as cited in Ibid., p. 36.
6. Kapoor, n. 1, p. 37.
7. Ibid.
8. Ibid., pp. 38-39, 42.
9. Carey Sublette, "Pakistan's Nuclear Weapons Program: The Beginning", at http://nuclearweaponarchive.org/Pakistan/PakOrigin.html
10. I.H. Usmani, "Atomic Energy in Pakistan", *Pakistan Quarterly*, Special Science Number, 10, No. 2, Autumn, 1960, p. 136, as cited in Dr Savita Dutt, *To Chagai and Beyond* (New Delhi: I.K. International Pvt. Ltd., 2003), p. 6.
11. Shahid-ur-Rehman, "A Tale of Two Scientists" *Long Road to Chagai* (Islamabad: Printwise Publication, 1999), p. 21 and 31, as cited in " Pakistan Profile: Nuclear Chronology 1953-1970", at http://www.defence.pk/forums/general-defence/13739-nuclear-chronology-pakistan.html
12. Kapur, n. 1, p. 54.
13. Raj Chengappa, *Weapons of Peace* (Delhi: Harper Collins, 2000), p. 93.
14. Ibid., p. 93.
15. Bharat Karnad, *India's Nuclear Policy* (London: Praeger Security International, 2008), p. 48.
16. Ibid.
17. Chengappa, n. 13, p. 94.
18. Zulfiqar Ali Bhutto, *The Myth of Independence* (London: Oxford University Press, 1969), p. 152.
19. Ibid., p. 153.
20. Kapur, n. 1, pp. 117-119.
21. June 6, 1968, cited in SIPRI, *Near-Nuclear Countries and the NPT,* p. 26, as cited in Kapur, n.1 p. 118.
22. This proposal was made in September 1968, cited in "Near-Nuclear Countries and the NPT," pp. 26-27, as cited in Kapur, n.1 pp.118-119.
23. Steve Weissman and Herbert Krosney, *The Islamic Bomb* (New York: Times Books, 1981), p. 25.
24. Zulfiqar Ali Bhutto, *If I Am Assassinated.....* (New Delhi: Vikas Publishing House, 1979), p. 137.

25. "Timeline of Pakistan's Nuclear Programme", *Pakdef.info*, at http://www.pakdef.info/nuclear&missile/timeline2.html
26. Bhutto, n. 24, p. 137.
27. "Pakistan Institute of Nuclear Science and Technology", http://www.paec.gov.pk/pinstech/
28. Bhutto, n. 18, p. 152.
29. Ibid., p. 153.
30. Weissman and Krosney, n. 23, pp. 17-18.
31. Owen Benett Jones, *Pakistan: Eye of the Storm* (New Delhi: Penguin Books, 2001), p. 196.
32. An account of the Multan meeting can be found in Rehman, n. 11, pp.16-19. Also see Zahid Malik, *Dr A.Q. Khan and The Islamic Bomb* (Islamabad: Hurmat Publication, 1992), p. 128, as cited in Ibid., p. 196.
33. Maj Gen D.K. Palit and P.K.S. Namboodiri, *Pakistan's Islamic Bomb* (New Delhi: Vikas Publishing House, 1979), p. 6.
34. Ibid., pp. 6-7.
35. Douglas Frantz and Catherine Collins, *The Man From Pakistan: The True Story of the World's Most Dangerous Smuggler* (New York: Twelve, 2007), p. 22.
36. Ibid., p. 22.
37. Ibid., p. 22.
38. Weissman and Krosney, n. 23, p. 67.
39. Bhutto, n. 24, p. 138.
40. Benett Jones, n. 31, pp. 197-198.
41. Weissman and Krosney, n. 23, pp. 187.
42. "Nuclear Black Markets: Pakistan, A.Q. Khan and the Rise of Proliferation Networks A Net Assessment", *IISS Strategic Dossier,* 2007, p. 17.
43. Weissman and Krosney, n. 23, p. 190.
44. Ibid., p. 192.
45. Frantz and Collins, n. 35, p. 31.
46. Weissman and Krosney, n. 23, p. 198.
47. Dutt, n.10, p. 30.
48. Ibid.
49. Weissman and Krosney, n. 23, p. 167.
50. Kapur, n. 1, p. 185.
51. Dutt, n. 10, p. 40.
52. Sublette, n. 9.
53. See declassified 'NODIS' (No Distribution) State Department telegrams published by the National Security Archive, at "US and British Combined to Delay Pakistani Nuclear Weapons Program in 1078-1981, Declassified Documents Show" at http://

www.wilsoncentre.org./publications/

54. Ibid.
55. Gen Khalid Mahmud Arif, *Working with Zia: Pakistan's Power Politics 1977-1988* (Karachi: Oxford University Press, 1995), pp. 340-341.
56. Tim Weiner, "Nuclear Anxiety: The Know-How; US and China Helped Pakistan Build Its Bomb", *The New York Times*, June 1, 1998.
57. Indira Gandhi, *Prime Minister Indira Gandhi: Statements on Foreign Policy: April-September 1982* (New Delhi: Ministry of Foreign Affairs, 1982), p. 9.
58. Weiner, n. 56.
59. William Burr, "The China-Pakistan Nuclear Connection Revealed", November 18, 2009, at http://nsarchive.wordpress.com/2009/11/18/the-china-pakistan-nuclear-connection-revealed/
60. S.Y. Khan, 1986, as cited in Syed Shahid Hussain Bukhari and Atiq- ur- Rehman, "Pakistan-China Nuclear Deal and International Fictions", *Berkeley Journal of Social Sciences*, Vol. 1, Issue 3, March 2011
61. Cited in Ibid.
62. US State Department, "The Pakistani Nuclear Program," June 22, 1983, secret, excised copy. Source: State Department FOIA release, See William Burr, ed., *China, Pakistan, and the Bomb: The Declassified File on US Policy, 1977-1997*, National Security Archive Electronic Briefing Book No. 114, March 5, 2004, at http://www.gwu.edu/~nsarchiv/NSAEBB/NSAEBB114/index.htm
63. Bill Gertz, *Washington Times*, October 4, 1994, p. A8; Barbara Starr, *Jane's Defence Weekly*, October 15, 1994, p. 6 as cited in "China's Missile Exports and Assistance to Pakistan—Statements and Developments" at http://cns.miis.edu/archive/country_india/china/mpakchr.htm
64. "Missile Overview", *NTI* at http://www.nti.org/e_research/profiles/Pakistan/Missile/index_3066.html
65. Dutt, n . 10, p. 50.
66. Ibid.
67. Kapur, n. 1, p. 210.
68. Zahid Hussain, *Frontline Pakistan: Path to Catastrophe and the Killing of Benazir Bhutto* (New Delhi: Penguin Group, 2008), p. 159.
69. See, Kuldip Nayar, "We Have the Bomb, Says Pakistan's Dr. Strangelove", *The Observer* (London), March 1, 1987.
70. Gen Mirza Aslam Beg, "Nuclear Programme and Political Ramblings", *Defence Journal*, January 1994-February 1994, pp. 19-21.
71. Ibid., pp. 19-21.
72. Ibid.
73. Gen Aslam Beg, *Nation*, December 13, 1993, in Jasjit Singh, *Kargil 1999* (New

Delhi: KW Publishers), p. 53.

74. Hans M. Kristensen and Robert S. Norris, "Pakistan's Nuclear Forces, 2011", *Bulletin of the Atomic Scientists,* 2011 67:91, p. 92.
75. Ibid.
76. http://www.fas.org/nuke/guide/pakistan/missile/hatf-3.htm, for details on Chinese transfer of missile technology to Pakistan. See, Chapter 3 – China-Pakistan Strategic Nexus: Strenthening the Conventional and Nuclear Capability.
77. "Pakistan's Nuclear Weapons Programme Development", at http://nuclearweapon archive.org/Pakistan/PakDevelop.html
78. "China Missile Exports and Assistance to Pakistan", *NTI*, at http://www.nti.org/db/china/mpakpos.htm
79. "Bhutto Ends Visit to North Korea", Agence France Presse, December 30, 1993, International News; in Lexis-Nexis Academic Universe, December 30, 1993, web. lexis-nexis.com, at "NTI: Country Overview: Pakistan Missile Chronology" at, http://www.nti.org/e_research/profiles/Pakistan/Missile/chronology_1992_1993.html
80. n.77.
81. Ibid.
82. For details, see Dianne E. Rennack, "India and Pakistan: Current US Economic Sanctions" , *CRS Report for Congress*, October 12, 2001.
83. For details, see Chapter 4 – Nuclear Delivery Systems.
84. The details for the status of Nuclear Power in Pakistan are taken from "Nuclear Power in Pakistan" at http://www.world-nuclear.org/info/inf108.html
85. Kristensen and Norris, n. 74, p. 91.

# 2. The Nuclear Bazaar: Exports to Iran, North Korea and Libya

One of the most critical aspects of Pakistan's clandestine nuclearisation has been the proliferation of nuclear technology to other countries. This has invariably posed severe challenges to global security and also the global disarmament efforts. Pakistan acquired nuclear weapon technology secretly and, subsequently, it used the same interlinked channels to supply the technology to Iran, North Korea and Libya. While there are other countries, including Saudi Arabia and Iraq, which have reportedly been beneficiaries of Pakistan's nuclear proliferation, channelised through the "Khan network", these three countries (Iran, North Korea and Libya) provide clear evidence of the wide scale technology transfer, including, equipment, blueprints, plans, training and, very importantly, the list of the sellers of critical parts.[1]

Pakistani state authorities have consistently denied involvement in the transfer of nuclear technology to other nations and have managed to wash their hands off any accountability by putting the entire blame on one man, Dr A. Q. Khan and his associates for the proliferation. However, it is difficult to believe that in a country where the military leadership always commanded strict control over the nuclear arsenal, to the extent that even in the 1990s when the military was not directly in power, the nuclear programme was strictly guarded by the army, one man could export the technology out to other interested nations, through a network which operated across as many as twelve countries, over more than a decade. In fact, Pakistan was caught in a dilemma after the detection of technology transfer to various countries: it was trying to make a distinction between the state's nuclear weapon programme and a group of scientists who, it claimed, plunged into this massive international trade for personal gains. Pakistan's proliferation shows a lack of responsibility in terms of handling the nuclear technology and a failure of the state command and control over its nuclear arsenal. And a satisfactory explanation for these issues is still lacking.

A. Q. Khan himself boasted, "A country, which could not make sewing needles, good bicycles or even ordinary durable metalled roads was embarking on one of the latest and most difficult technologies".[2] Khan took immense pride in being the father of the Pakistani nuclear programme and held the view that the technology should be passed on to other nations as well. The Pakistani leaders have denied any knowledge of the Khan network. But various reports suggest that Khan and his associates were very vocal about advertising the technology they acquired in the 1980s. The Khan Research Laboratory (KRL) was circulating a brochure with the print line "Government of Pakistan" along with A. Q. Khan's picture on the cover of the brochure to the states seeking nuclear technology.[3] They were openly displaying their willingness to sell the technology in various international arms fairs.

Gen Musharraf was well aware of the fact that Khan was involved in illicit trade (nuclear) dealings and also, Khan's financial assets were no secret. Musharraf was under severe pressure from the US to take action and he removed Khan from his office in 2001. Khan was pardoned after his admission and he retired with full dignity and honour. Gen Musharraf, in his speech at Dr Khan's retirement dinner on March 27, 2001, said:

> The situation was critical. Our security paradigm had changes and, with no nuclear weapons programme worth the name, Pakistanis literally looked to the sky for help. We did not lose faith. And sure enough, Allah Almighty answered the nation's prayers, had mercy on our situation and made a miracle happen. In walked a giant of a man, none other than Dr Abdul Quadeer Khan, the man who would give Pakistan a nuclear capability single-handedly.[4]

In 2004, after revelations of the US intelligence agencies, Dr A. Q. Khan admitted having transferred nuclear technology to North Korea, Iran and Libya. He reportedly accepted the allegation about transfer of nuclear technology to the three countries in a signed 12-page document.[5] The Pakistani government called it a lapse in intelligence and a failure of the system to safeguard the nuclear technology. According to the media reports within Pakistan, the stance of the Pakistani authorities was: "Everybody knew ours was a covert programme and every successive government and security agency overlooked allegations about Dr Khan's assets in the interest of the programme and because of the trust in his person, who, without any doubt, was a towering personality."[6]

Dr Khan was pardoned by Gen Musharraf given his immense contribution to the nuclear programme. After the disclosure of the proliferation of nuclear technology from Pakistan, almost a two-month-long debriefing of the top scientists and officials of the KRL was carried out which apparently produced sufficient evidence of nuclear proliferation.[7]

The transfer of nuclear technology/equipment from Pakistan highlighted an extremely complex trend in proliferation, where the sellers and buyers were spread out in more than a dozen countries. It also brought out the various loopholes in the export control regime. However, to understand Pakistan's objectives and the nature of its involvement in the proliferation, it would be useful to study its nuclear export to three countries – Iran, North Korea and Libya.

**Pakistan and Iran**

The suspicion regarding Iran's nuclear programme goes back to the 1980s and early 1990s. In 2002, a London-based Iranian opposition group disclosed the facts regarding the secret heavy-water production plant at Arak, and the underground enrichment facility at Natanz.[8] Reports also suggested that in 2002, satellite photographs had revealed undeclared nuclear facilities under construction.[9] These revelations were followed by Iran's admission wherein it conceded aspects of its nuclear programme, which intensified the International Atomic Energy Agency (IAEA) inspections.

In December 2003, *The Washington Post* reported:

> Evidence discovered in a probe of Iran's secret nuclear program points overwhelmingly to Pakistan as the source of crucial technology that put Iran on a fast track toward becoming a nuclear weapon power, according to US and European officials familiar with the investigation.[10]

The documents provided by Iran to UN nuclear inspectors in November 2003, exposed a vast, clandestine procurement network which was spread into many countries for procuring different parts from different nations over a period of 17 years.[11]

*The Washington Post* reported:

> While Iran has not directly identified Pakistan as a supplier, Pakistani individuals and companies are strongly implicated as sources of key blueprints, technical guidance and equipment for a pilot uranium-enrichment plant that was first exposed by Iranian dissidents 18 months ago, government officials and independent weapons experts said.[12]

Investigations disclosed crucial assistance from Pakistan, and the blueprint which the IAEA reviewed, showed a type of centrifuge that was identical to a machine used by Pakistan in the early years of its nuclear programme. Pakistan offered the plans and components to Iran in the 1980s, and in the 1990s, provided a more advanced centrifuge design, which eventually allowed Iran to overcome the technical hurdles and make its own enriched uranium, which is an essential component in commercial nuclear fuel and a nuclear weapon. [13]

***Pakistan's Role in Iranian Nuclear Programme***

Iran's nuclear programme dates back to the late 1950s, and in 1957, Washington and Tehran signed a civil nuclear cooperation agreement under the US "Atoms for Peace" programme. Iran received from the United States a 5-Megawatt-Thermal (MWt) research reactor in 1959, and in 1967, the United States supplied Iran with 5.5 kg of enriched uranium and 112 grams of plutonium.[14] Iran under the Shah, before the 1979 Iranian revolution, was keen on moving ahead with the nuclear energy programme. In 1976, Iran signed a deal with a German company to build a nuclear reactor in Bushehr. During these early years of Iranian nuclear activity, the Shah's main concern was energy. Iran's inclination for nuclear technology encouraged the Iranian youth to go abroad for technical education.[15] For the Shah, the nuclear programme was a matter of prestige and a means of restoration of the Persian prestige. The United States, while encouraging and helping the Shah with nuclear reactors, was always suspicious regarding the Iranian nuclear programme and the intelligence agencies suspected that the Shah might be pursuing a clandestine nuclear weapon.[16] But after the Iranian revolution in 1979 and the overthrow of the Shah , Iran's new leader Ayatollah Khamenei had a different opinion and regarded the nuclear weapons as *unIslamic*. The Shah's efforts went in vain and the nuclear programme was pushed back. It was after the Iraqi invasion in 1980, when Iraq deployed chemical weapons, that Iran not only started to develop its own chemical weapon programme but also took a relook at the nuclear option. Iranian President Hashemi Rafsanjani started to explore possible foreign assistance for the nuclear programme. China and Russia have been the known sources of nuclear and missile assistance to Iran. But during the 1990s, there was significant pressure from the US on China and Russia to prevent transfer of nuclear and missile technology to Iran.

Iran's relationship with Pakistan had improved in the mid-1980s during Zia's time as Pakistan started to import oil from Iran. Iranian President Khamenei visited Pakistan in 1986 and diplomatic contacts between the two countries were established. Reportedly, Zia had given the nod for civil nuclear cooperation but was cautious and did not allow the transfer of any sensitive technology to Iran. According to Pakistani officials, Zia did not allow any agreement which would involve sensitive technology, partially because of the American monitoring since Islamabad did not want to compromise its own programme at any cost, and, secondly, Zia's resistance was also due to his strong Sunni beliefs which put him in a difficult situation in his dealings with Khamenei.[17]

There is a lack of clarity as to when the first contact between Dr Khan and Iran was established. Some reports suggest that it was as early as 1985 when a Pakistani delegation [probably followed by Inter-Services Intelligence (ISI) Chief Gen Hamid Gul in 1987] held its first round of talks (possibly) selling nuclear technology to the representatives of Muslim countries, including Iran and Libya, and that is when Dr Khan's association with Iran started. In 1987, Iran entered into a formal civil nuclear cooperation agreement with Pakistan. The deal apparently included the training of at least six Iranian scientists by the Pakistani scientists at PINSTECH. In the same period, Dr Khan paid a visit to Iran to have a look at the Bushehr reactor in order to assess the damage caused by the Iraqi bombing.[18]

Iran was keen on buying the technology and Khan was ready to sell. Although, the details of who initiated the clandestine cooperation (Khan or the Iranians) are not clear, the first crucial meeting was held in Dubai in 1987, in which the key members of the Khan network were present. The Iranians reportedly met S. Mohamed Farouq, head of the computer import-export company SMB Group and his nephew B. S. Tahir who played a key role and became more or less Dr Khan's acting agents in marketing the P-1 centrifuge components.

In 2005, international investigations disclosed the details of the Dubai meeting that resulted in a written offer to supply Tehran with the makings of a nuclear weapon programme.[19] The IAEA probe and monitoring did manage to obtain the copy of the offer which was made to Iran by Pakistan. Apparently, the offer was described as a five-point, phased plan in which the Khan network offered to supply Iran with "drawings for Pakistani centrifuges and then a starter kit of one or two centrifuges. Phase three included as many as 2,000 centrifuges, which could be used to enrich bomb-grade uranium. Auxiliary items for the centrifuges and enrichment process would have been delivered afterward, followed by reconversion and casting equipment for building the core of a bomb."[20] The deal was worth three million dollars. Khan also offered the Iranians a list of what to buy and from where to buy. Iran bought some equipment from the Khan network and it acquired quite a bit of equipment from the European countries.

Iran was determined to get the technology and in 1988, when Iraq used chemical weapons, Rafsanjani told the Iranian Parliament: "We should at least think about [weapons of mass destruction] for our own defence. Even if the use of such weapons is inhuman and illegal, the war has taught us that such laws are just drops of ink on paper."[21]

After the deaths of Zia and Khomeini in 1988 and 1989 respectively, the Iranian inclination for nuclear technology increased and, on the other side, the new military leadership in Pakistan under the Army Chief Gen Aslam Beg supported the Iranian cause and was in favour of Pakistan extending cooperation with Iran. This coincided with Pakistan ending its alliance with the Americans.

Aslam Beg encouraged assistance to Iran; although he has denied his involvement in the transfer of nuclear technology, he has admitted that there were frequent nuclear discussions in the early 1990s between Iran and Pakistan. Ghulam Ishaq Khan, Zia's successor, provided full protection to Khan and was always in favour of providing nuclear support to Iran. Also, Benazir was aware that during both her terms in the 1990s, talks with Iran on nuclear cooperation were going on.[22] It is difficult to come to a conclusion about who directed Khan, but a major deal was struck in 1994, which included P-1 designs and drawings for advanced P-2 centrifuges.[23] The centrifuges and components provided by Khan were mostly second-hand cast-offs from Khan's warehouse in Dubai, and Iran received equipment which was not needed by Pakistan. As a result, the Iranians had frequent meetings with Khan and his associates, most of which reportedly were due to complaints regarding the equipment.[24] Iran began to diversify its procurements and moved towards Turkey and Eastern Europe. There are reports suggesting that Khan also supplied P-2 centrifuges to Iran in the mid-1990s. Although, Iran was procuring equipment from Europe to evade export controls, the fact is that it was the Pakistani assistance which got the Iranian nuclear programme started. Iran received all the crucial inputs from Pakistan—equipment, advice, technology, etc. The Iranian clandestine enrichment programme stepped up from 1995 and the Kalaye Electric Company came up in the suburb of Tehran, located in a small building. The Iranians claimed that it was a clock factory, but investigations revealed that the P-1 centrifuges were being produced here. Also, here the uranium hexafluoride gas (UF6) was fed into a centrifuge to be enriched in 1999. And by 2000, Iran was prepared to start the construction of the two enrichment facilities at Natanz.[25]

Natanz was built in deep bunkers reinforced by triple layers and then covered with earth to hide their existence, as Jordon Corera points out. Without going into the details of the Iranian nuclear programme, it would be apt to say that it went on at a good pace and in 2006, Iranian President Ahmadinejad declared that Iran had enriched uranium for the first time, and later, in 2007, Iran declared that it had 3,000 centrifuges enriching uranium.

Pakistan's role in Iran's civil nuclear programme and for the weapon technology is very obvious. The wide ranging cooperation between the two countries increased after Zia's death. Gen Aslam Beg was in favour of promoting a strategic triangle comprising Pakistan, Iran and Turkey, (originally promoted by Z. A. Bhutto as the ECO) to oppose the US dominance in the Middle East. Dr Khan certainly had a strong financial incentive in dealing with Iran, apart from the ideological factor to pass on the technology to another Muslim nation. Khan, and also the military leadership in Pakistan believed that since Pakistan had the technology, other Muslim nations should also possess it. Khan believed that the other Islamic countries should acquire nuclear capability in order to ease the intense Western pressure on Pakistan's nuclear power.[26]

It is possible that the transfer of nuclear weapon technology did not have the full concurrence of Pakistan's political leadership, as there are reports of Benazir not giving approval for the weapon technology transfer, but it is fair to assume that Benazir and later, Nawaz Sharif, were aware of the transaction. The political leadership in Pakistan, till today, does not have the leverage to interfere in the overall defence matters, let alone nuclear issues. Given this, it is possible that Khan had the go-ahead from the military leadership, without consulting the civilian leaders.

**Pakistan and North Korea**

Pakistan was a major supplier of critical equipment for the North Korean nuclear programme. Despite signing an agreement to abandon its research efforts in 1994, North Korea admitted that it was pursuing a nuclear weapon programme. In October 2002, the Bush Administration announced that North Korea had been pursuing a clandestine uranium enrichment programme. After a few days, the US press reported the involvement of Pakistan in the North Korean efforts.[27] It was as early as 2000, when the US intelligence detected evidence of North Korea's clandestine efforts to procure high-strength aluminium tubes to build gas-centrifuges. The disclosure of these facts, and also other undisclosed evidence led the US to conclude that Pakistan was the source of the North Korean uranium enrichment technology. [28] This conclusion was derived partly due to the imagery analysis of the unmarked containers loaded on Pakistani C-130 transport aircraft. [29]Although the details of the technical transaction between the two nations are not given precisely, the US intelligence reports did confirm the exchange of scientific personnel and also "highly questionable" shipments to North Korea.[30] The reports convincingly suggested that Pakistan had traded the uranium technology in return for the North Korean No-dong missiles. In 2004, Dr A. Q. Khan, after his arrest, confessed having transferred centrifuges and related technology to North Korea, in the late 1990s.[31] Later, there were other accounts which were made public, bringing Pakistan's nuclear exports into the limelight. Gen Musharraf's autobiography, *In the Line of Fire: A Memoir*, published in 2006, gave a much more detailed description of the Pakistan-North Korean transaction.

***Pakistan-North Korea: Nuclear-Missile Links***

Pakistan's relationship with North Korea dates back to Zulfiqar Ali Bhutto's regime in the 1970s. Pakistan had aggressively started to explore alternate options for weapon procurement after US sanctions were imposed in 1965. Pakistan's natural inclination was towards the Chinese equipment as the strategic partnership between Pakistan and China had been formalised with the signing of the Shaksgam Valley agreement in 1963 and also, China had offered its support to Pakistan during the 1965 India-Pakistan War. Chinese weapons were also cheaper and the Chinese obviously encouraged supplies to Pakistan by offering loans on easy repayment basis. North Korean equipment was also much more affordable for Pakistan and in the mid-1970s, Bhutto approached North Korea for assistance in replenishing Pakistan's depleted stock of military equipment. He established diplomatic relations with North Korea in 1976, but the ties between the two countries were frozen in the late 1970s when Zia-ul-Haq took over Pakistan. Zia's regime coincided with the Soviet invasion of Afghanistan and Pakistan's alliance with the United States.

Pakistan managed to acquire the nuclear weapon capability by 1987 and was capable of building uranium enriched devices. Zia negotiated the F-16s from the US as the primary delivery mechanism and managed to get 40 F-16s. In 1990, the alliance with the US came to an end and the US supplies were suspended, with the sanctions imposed under the Pressler Amendment. Pakistan was unable to receive even the 28 F-16s it had already paid for. It turned towards the option of a ballistic missile force in order to build credible nuclear strike capability against India. Pakistan's alliance with China in the field of defence cooperation was already in line. China had been a major partner in Pakistan's nuclear build-up and, not only had Pakistan acquired the Chinese equipment in the 1970s and 1980s but the Chinese had also started to invest in Pakistan's defence industry.

Pakistan and China reportedly started discussing the transfer of M-11 missile technology to Pakistan in the late 1980s and the contract for the sale was signed in 1987. This was followed by the US sanctions against China.[32] China had pledged in 1991 and 1994 that it would comply with the Missile Technology Control Regime (MTCR) and would abstain from missile technology transfer. But, in December 1992, reports suggested that China had apparently transferred 34 complete M-11 systems to Pakistan. These reports were denied by China.[33] Although Pakistani leaders have made conflicting statements regarding the receipt of the missiles, various reports provide enough evidence that Pakistan did receive the M-9 and M-11 missiles from China. China also helped Pakistan to build the M-11 factory near Rawalpindi.

The US administration took measures to dissuade China from transferring the missiles, components and missile related technology to Pakistan, and was engaged in a series of negotiations seeking to obtain pledges by China to abide by the MTCR guidelines (formulated in 1987) and also to become a member of the regime.[34] Washington offered economic incentives, took diplomatic measures and also imposed economic sanctions on China. Beijing made pledges in 1991 and 1994 when it signed the 31-nation MTCR as well as a 1994 US-China agreement not to deploy M-11s in Pakistan.[35] After Washington received the reports of the transfer of the M-11s to Pakistan, limited sanctions were imposed on China in 1993.[36] China's missile transfer was closely monitored by the US, thus, it was logical for Pakistan to explore a different option which could provide it with the Chinese technology or equally suitable technology with a longer range missile.

Pakistan explored the No-dong system with North Korea as early as 1992, and Pakistani officials visited North Korea to view a No-dong prototype. In May 1993, Pakistani scientists and engineers visited North Korea and attended the No-dong test launch.[37]

Later, in December 1993, Prime Minister Benazir Bhutto visited Pyongyang after having talks with North Korean President Kim Il Sung. Benazir received an extremely warm welcome in Pyongyang and during her stay, she managed to strike a critical deal for Pakistan. Reportedly, it was A. Q. Khan who was instrumental in organising Benazir's visit to Pyongyang—Khan would supposed to be at the centre of the nuclear relationship between the two countries.[38] Jordon Corera, in his detailed study on Pakistan's proliferation says: "North Korea is one of the most secretive nations in the world and its program – and dealings with Khan – are shrouded in layer upon layer of mystery and speculation."[39]

In late 1995, Marshal Ch'oe Gwang, the former Vice-Chairman of North Korea's National Defence Commission, visited Pakistan and brokered the deal. Later, in 1996, Taiwanese officials seized 15 tons of ammonium perchlorate, an oxidising agent used in modern solid-propellant formulas, on a freighter bound from North Korea to Pakistan's Space and Upper Atmosphere Research Committee (SUPARCO).[40] In 1997, Kang T'ae Yun, a North Korean diplomat based in Pakistan, arranged for the supply of maraging steel from the All Russian Institute of Alloys in Moscow, to North Korea and Pakistan.[41] The foreign intelligence agencies started monitoring the flights between North Korea and Pakistan and, reportedly, the frequency of the flights increased from nearly three a month in autumn 1997 to about three times that number in January 1998. The flights carried the technical experts, including the North Korean telemetry crews.[42] Reportedly, the 5-10 No-dong missiles assembly sets were sent to Pakistan between 1994 to 1997 for trials and to set up a manufacturing capability.[43] Pakistan's Ghauri or Hatf -5 missile that resembles the Russian 'Scud B', is an outcome of coordinated inputs from both North Korea and China. It has been reported that China provided Pakistan with a nuclear warhead design with a weight of 500 kg in the early 1980s, and this design was used to provide the warhead for the Ghauri missile. Also, there have been reports that the Chinese guidance systems have been used, which could have been passed onto Pakistan through North Korea.[44] The first flight test of the Ghauri, single stage, liquid propellant missile, with a range of 800-1,200 km, was in April 1998. Testing the Ghauri was an important moment for Pakistan as according to the Pakistani leadership, it would give them an advantage over India's strategic assets as some of the important Indian cities, power plants, water purifiers would fall in this range.[45]

Transfer of the North Korean No-dong missile raised obvious questions regarding what North Korea had gained in return. It has been estimated that the cost of the No-dongs would have been approximately three billion dollars, if not more.[46] By the late 1990s, it was getting harder for Pakistan to pay North Korea in cash as the Pakistani economy was in the doldrums, with declining foreign reserves and rising debt. The GDP growth rate also went down to as low as 1.9 percent in 1997. And, thus, Pakistan could have potentially agreed for a different mode of payment – enrichment technology. The exchange between the two countries worked out very well as North Korea was looking for assistance with its clandestine enrichment programme in the mid-1990s. After the US had frozen North Korea's plutonium-production facilities in Yongbyon, it was desperate to build nuclear weapons clandestinely. North Korea believed that it had to get the bomb and was ready to explore every option.[47] While it is unclear when the North Korean enrichment programme began, there are reports about North Korea's procurement of enrichment related equipment in the mid-1980s.[48]

A. Q. Khan admitted that he started dealing with North Korea regarding enrichment in the late 1980s, although the major supplies went only in the late 1990s. Another important source revealing the transfer of technology in return for missiles came from Hwang Jang-yop, a Communist Party Secretary, who stated that in the summer of 1996, Pakistan agreed to trade North Korean long-range missile technology for enrichment technology. [49]

The trade of the enrichment technology suited financially pressed Pakistan in the late 1990s. At the same time, reports suggest that there was significant increase in the number of transport aircraft which visited North Korea from Pakistan. By 1998, there were nine flights per month, and frequent visits of high level North Korean officials to Pakistan.[50] In December 1997, Pakistan Army Chief Jehangir Karamat visited

Pyongyang and although there are no direct signs, roughly this was the time when transfer of enrichment technology also commenced. North Korea sent three engineers to Pakistan in 1999 to study uranium enrichment technology. The three engineers went to the nuclear institute while A. Q. Khan was heading the institute.[51] Khan himself travelled approximately thirteen times to North Korea from 1997 and his planes were arriving in North Korea till 2002.[52] Although the Pakistani state authorities denied that they had any knowledge of Khan's activities, and the fact that Khan was actually transferring the uranium technology to North Korea, later, President Musharraf, after years of denial, admitted that "probably a dozen" centrifuges were sent to North Korea during the late 1990s and early 2000. In some places, Musharraf mentioned the figure of twenty.[53] Musharraf said that Khan had sent uranium hexafluoride to North Korea and also designs for facilities and, more importantly, the designs and the blueprint were also passed on to North Korea. While the transfer of uranium took place between 1997 and 1999, the technical advice and guidance from Khan in terms of the important items to buy, and from where to buy was provided by Khan later.[54] Musharraf himself wrote about the North Korean dealings with Khan: "Some North Korean nuclear experts, under the guise of missile engineers, had arrived at KRL and were being given secret briefings on centrifuges, including some visits to the plant."[55]

North Korea tried to obtain material from other sources, including China, Japan, Europe and Russia but Pakistan provided most of the assistance related to the centrifuge rotors.[56] Musharraf, in his memoir, has disclosed that nearly two dozen P-1 and P-II centrifuges were transferred to North Korea. Apart from this, "flow meters, some special oils for centrifuges and coaching on centrifuge technology, including visits to top-secret centrifuge plants" were provided to the North Koreans.[57] The Pakistan-North Korean missile-uranium technology relationship stoked various speculations and it is believed that not only were the North Koreans present at the time of Pakistan's nuclear tests in 1998, but also that Pakistan may have tested a North Korean device in addition to its own.[58]

The nuclear crisis involving North Korea had erupted in 2002 when Pyongyang admitted having a covert weapons programme. The Pakistani state authorities have been in consistent denial of the involvement of the state in the transfer of technology to North Korea and some analysts have suggested that nuclear cooperation with North Korea was a clandestine operation conducted by A. Q. Khan and KRL and had no state authorisation. But the fact that Khan and KRL could have worked out a deal with the North Koreans independently, seems implausible. The Pakistani nuclear programme has always been under the tight control of the army (after Zia–ul-Haq took over) and even the ruling democratic governments had authority to intervene in the nuclear matters. Khan's visits were well known in the Pakistani military and intelligence community and another critical question relates to how Khan managed to transfer the centrifuges on military planes without the knowledge of the military.[59] Khan in his confessions is reported to have said that the Chiefs of Army Staff—Gen Karamat, Gen Waheed and Gen Musharraf—knew of the missile dealings and were well aware of the nuclear technology which was traded in return for the No-dongs. The army, on the other hand, has maintained the stance that the North Korean deal was Khan's personal initiative and did not have state support.

**Pakistan and Libya**

The Libyan deal was very different from those with Iran and North Korea. These countries had a fair amount of technical expertise but required assistance. However, in the case of Libya, this was not so. The Libyans had to be provided everything from scratch – the entire nuclear capability.

In October 2003, the US spotted a German-registered ship bound for Libya through the Suez Canal, with thousands of parts of uranium centrifuges.[60] The shipment came from Dubai which was the hub of Khan's secret activities. The heat of US counter-proliferation had caught up and this obviously pressurised Libya in December 2003 to admit that it did have a nuclear weapon programme and also an assurance from the Libyan leader Col. Muammar-al-Gaddafi to put a stop to it. Libya was still far away from producing a nuclear weapon and important pieces of equipment were eventually dismantled and stored in boxes, as claimed by the top United Nations nuclear inspector, Dr. Mohamed El Baradei.[61] On his visit to Libya in December 2003, he said: "Luckily, I think we are where they have not developed a full-fledged capability, and luckily also, they have announced that they are ready to eliminate all programs relevant to weapons of mass destruction."[62]

Ties between Pakistan and Libya go back to the early 1970s, when Pakistan had just begun its efforts for the nuclear bomb. Libya signed a deal with Pakistan and agreed to help in financing Pakistan's atomic efforts in return for knowledge about how to make nuclear fuel.[63] It was also been reported that from 1978 to 1980, Libya may have supplied Pakistan with the uranium core.[64]

From the late 1980s, after Pakistan acquired nuclear weapons capability, Khan went across the Middle East and North Africa, talking to people to see if there were potential buyers for the nuclear technology. By the end of the 1990s, KRL was propagating its willingness openly in the various international arms fairs to sell not only conventional weapons but also centrifuge technology.[65] On the other hand, from the mid-1980s onwards, the Libyan nuclear programme was not progressing and Libya was frustrated as there was lack of willingness on the part of the nuclear suppliers to deal with the regime.[66] By 1995, Libya contacted Khan to conclude a deal for the uranium enrichment programme. Khan met the Libyans along with his loyal lieutenant B. S. Tahir. By 1997, Libya received the first tranche of equipment to begin its research, which consisted of 20 assembled P-1 centrifuges and parts for 200 more.[67] The Libyans were keen on getting an entire centrifuge facility, and in 2000, Pakistan sent Libya two test P-2 (renamed L-2) maraging steel centrifuges. Both these centrifuges have been used in the Pakistani nuclear programme.[68] Gaddafi had ordered 10,000 additional P-2 centrifuges, enough to produce fissile material for up to ten bombs a year. This order included designs for the plant, 20 tons of UF6 and most of the associated equipment. Regular consultations and training sessions (around the world) on the use of the equipment were also a part of the deal.[69] The first deliveries were made in 2002. Thus, the deal with Libya was significantly large for Pakistan, and the Libyans certainly would have given a lucrative offer to Pakistan. To fulfil the Libyan deal, expansion in the network was required and it is estimated that up to 30 companies in 12 countries would have been involved in this trade. [70] The personnel involved in the network had a history of selling or purchasing items for the Pakistani nuclear programme. Many of Khan's European and South African partners stood by him; the network was vastly scattered and different parts were being manufactured in different countries all over the globe and were finally assembled in Libya in order to escape detection by the Western intelligence agencies.

Dubai became the hub for all the activities related to the Libyan nuclear trade. The Dubai based companies which have been mentioned in the proliferation network were Aryash Trading Company and Gulf Technical Industries (GTI), which was managed by a British citizen, Peter Griffin.[71] Libya was receiving shipments of nuclear technology from all over the world. Electronic regulators and power supplies came from Turkey, sophisticated machine tools from Switzerland and South Korea, substantive quantities of high-strength aluminium and maraging steel from Malaysia and Singapore, the P-2 centrifuge prototype from Pakistan, and also two form-flowing lathes from Spain.[72] The investigators identified China as the origin of the nuclear weapon design, technical instructions and uranium hexafluoride to Libya.[73] Dozens of Libyan technicians were being trained in Dubai, Spain and other countries. The basics were arriving at Libya, although it was still not close to building a bomb. The components were designed somewhere else, manufactured somewhere else, and assembled somewhere else, to be finally used in Libya. The illicit nuclear trafficking had been going on for some time now and, thus, it had acquired the professional skill to conceal the equipment from the intelligence agencies. For example, some of the equipment had civilian application and was openly transported to Libya by reputed companies without knowing that it was for the nuclear weapon programme.[74] Tahir was the manager and the chief finance officer for Dr Khan. Most of the equipment and machinery was first shipped to his SMB Computers in Dubai and then dispatched for Libya after repacking. Tahir handled the payment (which ran into billions of dollars) for Europe and other countries for the supplies through multiple international banks and front companies. Although it is difficult to provide the exact estimates of the supplies, reportedly, Libya was charged a much higher rate—almost double of the ongoing market price—and by summer 2003, Libya's payments to the network were estimated to be around eighty million dollars.[75]

As the Libyan deal was widespread and global in nature, the obvious questions which were raised were on the global export control laws. The deal exposed many loopholes in the export control regimes. Most of the countries involved in the network enjoyed laxity in terms of what came and went out of the country. Malaysia, for example, is not a member of the Nuclear Suppliers Group (NSG) and has no export control laws, and, thus, it could easily export and import goods. Similarly, in Dubai, there are hardly any restrictions on imports and exports, especially in the free trade zones.[76]

Pakistan's deal with Libya was important because it had a much more widespread network involving numerous companies in many different countries. Also the deal with Libya took place after Khan was removed from KRL in 2001. Interestingly, although he was not in KRL, he sent centrifuge components on government-chartered cargo flights directly from Pakistan. He used government chartered C-130s to transfer enrichment technology to Libya.[77] This could not have been possible without the knowledge and support of the state authorities. The nuclear cooperation between Pakistan and Libya dates back to the 1970s. Pakistan's success in its nuclear programme in the 1980s encouraged and motivated Libya to go ahead.

**Pakistan's Position on the NPT and CTBT**

Pakistan is a non-signatory to the Non-Proliferation Treaty (NPT) and has for long maintained the position that its signing the NPT is conditional on India signing it. Before the 1960s, Pakistan position on arms control and disarmament was not completely negative and in 1951, in the sixth session of the General Assembly, it did make a call for total nuclear disarmament. In the mid-1950s, it signed a statute for the International Atomic Energy Agency. In 1963, Pakistan signed the Partial Test Ban Treaty but did not ratify it, pointing to the flaws in the treaty. Pakistan's stance was that the treaty prohibited nuclear tests in the atmosphere, outer space and underwater, but it failed to stop the nuclear states party to the treaty from further developing and refining their nuclear capabilities. Thus, Pakistan retained its nuclear option to test even though India signed the treaty and also ratified it.

The NPT was finalised in 1968 and was unacceptable to both India and Pakistan. India rejected the treaty on the grounds that it provided the Nuclear Weapon States (NWS) a discriminatory right to develop, and control the use of, the weapon. Pakistan declined signing the treaty but was not vocal in its approach regarding the fact that its positioning was subject to India's stance on the treaty, and its official statements did not mention the India factor. In the 1960s, Pakistan was keen on preventing India from acquiring nuclear weapon capability. In the 1970s, after Bhutto overtly declared the initiation of the weapon programme and managed to get the scientific community together for the "bomb", especially after India's peaceful test in 1974, Pakistan's position became even more aggressive and it refused to participate in the NPT Review Conference in 1975. Its argument was that the NPT has failed to prevent India from going ahead with the explosion. But the fact is that Pakistan had started building up its weapon programme even before 1974 and it saw the NPT as a hindrance to its nuclear programme; its position on the NPT was to protect its clandestine nuclear weapon programme from any external interference. Thus, in the 1970s, Pakistan took a rigid stand, was not in favour of allowing international inspection of its nuclear installations and made it clear that it was not prepared to accept any controls and inspections on its programme. In the 1980s, during its strategic alliance with the US, Gen Zia had clearly indicated to Washington that Pakistan's cooperation was conditional on the fact that America would not question Pakistan's nuclear weapon programme. Subsequently, Pakistan has maintained an overt position that it would sign the NPT if India signs it. The build-up of Pakistan's nuclear weapons is India-centric and to balance India's conventional military superiority. Hence, for Pakistan, signing the NPT was out of the question: nuclear weapons for Pakistan are a life saving device and the ultimate guarantor of its national security.

After the overt nuclearisation in 1998, Pakistan demanded its recognition as a nuclear weapon state as a condition for signing the NPT. Post the nuclear tests, the international community imposed economic sanctions on Pakistan. The G-8 countries as well the European Parliament insisted on Pakistan joining the NPT as a non-nuclear weapon state which Pakistan refused. Maintaining its position, Pakistan did not participate in the 2000 NPT Review Conference, which again called upon Pakistan to participate as a non-nuclear weapon state. Pakistani Foreign Ministry spokesman Abdul Basit said: "We cannot sign the treaty. We cannot give up nuclear weapons. If you have a conventional imbalance between Pakistan and India, then obviously our reliance on nuclear deterrence increases correspondingly."[78]

He added: "At one point in time, we were ready to sign the treaty, provided it was done by India. That is the position we took, but it has become outdated."[79]

Pakistan's reliance on nuclear deterrence has enhanced significantly in the last two decades and, thus, signing the NPT certainly does not fit into its strategic calculus.

Given Pakistan's position on the NPT, its decision not to sign the Comprehensive Test Ban Treaty (CTBT) did not come as a surprise. The CTBT was ready for signature in 1996, but Pakistan declined to sign it as the treaty has been perceived as an impediment to Pakistan's nuclear deterrent capability. Pakistan held the position that it would be unreasonable to demand its acceptance of the CTBT under Western pressures through sanctions. Keen on keeping the option of nuclear testing open, Pakistan made its signing of the CTBT conditional on India signing it. Post 1998 nuclear tests, the CTBT debate intensified both internationally and at the domestic level. Pakistan did not want to lose out on the option of further testing. In September 1999 Foreign Minister Sartaz Aziz said: "We will not consider the signing of the CTBT till the sanctions imposed against Pakistan by the United States are removed."[80]

Prime Minister Nawaz Sharif, while addressing the 53rd session of the United Nations General Assembly, made a statement: "In a nuclearized South Asia, CTBT would have relevance if Pakistan and India are both parties to the Treaty. However, Pakistan's adherence to the Treaty will take place only in conditions free from coercion or pressure."[81]

Pakistan has continued to maintain its non-adherence position on the CTBT. Pakistan's strategic calculus, the Indo-centricity of its nuclear weapons and the perceived role of nuclear weapons in dealing with its perceived implacable enemy have been major factors in determining its position on the CTBT. The perception is that signing the CTBT would confine the confrontation in the region to the conventional level, which would reduce Pakistan's deterrence capability. The domestic environment in Pakistan, including the political parties like the Jamaat-i-Islami and Muslim League and the public opinion, apart from the military leadership which controls the nuclear arsenal and policy, have opposed Pakistan signing the CTBT which is seen as 'suicidal' for Pakistan.[82]

On the Fissile Material Cut-off Treaty (FMCT), Pakistan has for many years not allowed the consensus decision to start the negotiations on the treaty at the Conference on Disarmament (CD) in Geneva. It adopted a dual contradictory approach: in one, to go along with the Islamic Middle East countries to question the existing stockpiles under the treaty (targeted at Israel), and in the other, with the rest of the international community to ban future production of fissile materials for weapons. Pakistan has been unsupportive of the FMCT, claiming that by calling for a ban on the production of fissile material while not taking into account the existing stockpiles, the FMCT ignores the nuclear asymmetry between India and Pakistan. Emphasising that India has a larger stockpile, Pakistan believes that the FMCT would assist in preserving the status-quo between the two countries.

Recently, Pakistan has raised another objection about the decision by the NSG to exempt India from the NSG restrictions on the sale of nuclear technology and material to countries out of the NPT. But it is not clear whether Pakistan would be willing to enter the FMCT talks in case it gets a similar waiver. Zamir Akram, Pakistan's Ambassador to the CD, said in an interview about Pakistan's position on the FMCT: "In the time that we can, we need to embrace our own capabilities so that we have sufficient fissile material for what we would then feel is a credible second-strike capability, or credible deterrence capability."[83]

The statement can be interpreted as meaning that Pakistan would slow down the talks on the FMCT and would potentially attempt to enhance its fissile material stocks with no assurances of finally signing the FMCT.[84]

**Some Conclusions**

The proliferation from Pakistan obviously created a massive uproar and also raised a number of questions regarding Pakistan's objectives and intentions behind the proliferation. It is not incorrect to say that export and proliferation of nuclear weapon technology from Pakistan has posed bigger challenges to global security than Al Qaeda. The worst nightmare of the West today is of nuclear weapons falling into the hands of *jihadis*. Coming to the question of why Pakistan indulged in this massive nuclear proliferation: there was no legal basis for Pakistan which could restrain it from proliferating nuclear weapons. While Pakistani state authorities continue to blame the proliferation on one man and his greed for monetary gains being the driving factor, this does not hold true. Some of the conclusions which can be drawn from the study of Pakistan's nuclear proliferation are discussed below.

Firstly, Pakistan's nuclear conduct from the beginning of its programme in the 1970s has been controversial. It acquired nuclear technology clandestinely, with A. Q. Khan as the main intermediary dealing with the transfers from Europe and China. Since the acquisition of the nuclear technology was done illegally, Islamabad saw nothing wrong in assisting other states to acquire it. The development of the nuclear programme in Pakistan was in a very different environment wherein it not only acquired the technology covertly, but also the nuclear organisations acquired immense powers granted by the then military leadership. Dr A. Q. Khan acquired immense power and respect from the state authorities, partly because of his personality and his desire to have that respect and authority, and mainly, because he managed to obtain quicker results as compared to his rival organisation, the Pakistan Atomic Energy Commission (PAEC).[85] Khan managed to move things much faster at Kahuta, and Gen Zia who particularly did not have a great liking for the PAEC Chairman, authorised Khan with major responsibilities. Khan was operating independently as the nuclear programme was to function secretly, given the Western pressures. This enhanced the rivalry between the PAEC Chairman and Khan which led to a lapse in the security apparatus. Thus, from the initial period of Pakistan's nuclear programme, there were hardly any restrictions on, and monitoring of, the technical transactions. Pakistan's involvement in the war against the Russians, along with the US, further created a fertile ground for the clandestine activities within the country as the nuclear related shipments were moving comfortably under the guise of shipments carrying weapons and other equipment. Islamabad went all out to create the base for the programme, and as A. Q. Khan said: "The world would never allow us to have it [nuclear weapons] and we would never bow to surrender it........We devised a strategy whereby we would go all out to buy everything that we needed in the open market to lay the foundation of a good infrastructure."[86]

Khan created a network of brokers, companies and financiers for Pakistan's nuclear programme which was now readily availbale to Islamabad for its nuclear exports.[87]

Secondly, like China, Pakistan felt that proliferation of nuclear weapon technology would be useful to create an alliance with like-minded nations in opposing the Western powers. Hence, it selected three radical, authoritarian and fundamentalist governments (Iran, Libya, Iraq) for this purpose. There has been some indication that Saudi Arabia was also to be included. Pakistan appears to have copied the strategy of Maoist China which sought to enlarge its following by giving nuclear technology (not necessarily up to the weapon stage) to the developing world to raise its own leverages against the US. Dr Khan himself resented the power and influence of the West, and for him, spreading nuclear technology to other states was a means of challenging the authority of the West.[88] According to Khan and other state parties involved, spreading the technology to other countries would also relieve the pressure on Pakistan regarding ownership of the weapon. Khan's bitterness for the West came out very explicitly in a letter he wrote to the German magazine *Der Spiegel*:

> I want to question the bloody holier-than-thou attitudes of the Americans and the British. These bastards are God-appointed guardians of the world to stockpile hundreds of thousands of nuclear warheads and have the God-given authority of carrying out explosions every month. But if we start a modest program, we are the Satans, the devils.[89]

Expanding Pakistan's influence in the Muslim world was certainly a contributing factor in the nuclear trade. Khan believed that "giving technology to a Muslim country was not a crime."[90] Also, he condemned the system that limited nuclear technology to the five major nuclear powers. He once said, "All Western countries are not only the enemies of Pakistan but, in fact, of Islam."[91] The proliferation of nuclear technology was not completely driven by the factor of Islamic solidarity, but it surely played a role in Pakistan's inclination towards these nations. Pakistan did manage to obtain funds to finance its nuclear programme from the wealthy Muslim states in the name of the *Islamic bomb* in the 1970s and later in the 1980s. Zulfiqar Ali Bhutto believed that Pakistan's stature as a nuclear weapon state would enhance its position and importance amongst the rich and security conscious Arab states.[92] Describing Pakistan's nuclear ambition in 1986, Zia said: "It is our right to obtain nuclear technology. And when we acquire this technology, the Islamic world shall possess it with us."[93] Khan, who was motivated to extend the nuclear technology to the Muslim states, openly admitted that he broke Pakistan's law by exporting technology but in doing so, he "acted in good faith". Whatever he did was to promote the "greater glory of Islam".[94]

Pakistan's nuclear status did enhance its image as the leader of the Islamic world. Given this background, Pakistan's exports to the Muslim world were not a surprise. The proliferation could be possibly motivated by feelings of ideological and religious solidarity with other Muslim or anti-American states. Apart from the strong factor of religion, these nations were in a position to offer good financial incentives to Pakistan.

Thirdly, it is significant to note that most of the countries to which Pakistan exported nuclear technology, controlled the bulk of the oil and gas reserves of the world. The likely aim of giving nuclear knowhow to some of these select regimes with large reserves of oil and gas, no doubt, also implied receiving favours in return, both for controlling energy supplies and for influence in the Islamic world.

Fourthly, Pakistan used its nuclear technology to obtain ballistic missiles (longer range). Initially, it acquired M-11 missiles from China but, subsequently, China came under tremendous US pressure to restrict its missile dealings. In due course, the dealings for missiles shifted to North Korea which itself had acquired missile technology from China.

Fifthly, Pakistan has widely claimed that this proliferation was undertaken by A. Q. Khan independently and essentially for commercial reasons. In a country in which the army has ruled and managed every policy, it is impossible to believe that in the most crucial area of national security policy like the nuclear weapons, this could be done by one scientist, for years, without the knowledge and approval of the army's top leadership. The successive army chiefs have repeatedly confirmed that the nuclear weapons have been under the tight control of the army. In fact, the army has ever strongly objected to the Chairman, Joint Chiefs of Staff being from the navy or air force. It is, therefore, unbelievable that this was an independent nuclear bazaar owned and managed by A. Q. Khan without the approval of the army leadership. On the other hand, since A. Q. Khan had sufficient experience in clandestine activities related to the nuclear programme, it is possible that he was an instrument of the proliferation policy at the behest of the army leadership.

### Notes

1. For additional information, see Peter Slevin, "At Least 7 Nations Tied to Pakistani Nuclear Ring", *Washington Post,* February 8, 2004.
2. Peter Edidin, "Pakistan's Hero: Dr Khan Got What He Wanted, and He Explains How," *The New York Times*, February 15, 2004.
3. David E. Sanger and William J. Broad, "Rogue Nuclear Projects: Tangle of Global Clues Has Pakistan at Center", *International Herald Tribune,* January 5, 2004.
4. Edidin, n. 2.
5. Nasir Iqbal, "Dr Khan 'Admits' He Transferred N-Technology: Action to be Decided by NCA," *Dawn,* February 2, 2004.
6. Ibid.
7. Ibid.
8. "Iran's Nuclear Program", Council on Foreign Relations, March 10, 2010, at http://www.cfr.org/iran/irans-nuclear-program/p16811
9. See, *Iran's Nuclear Program: Realities and Repercussions* (The Emirates Centre for Strategic Studies and Research, 2006).
10. Joby Warrick, "Nuclear Program in Iran Tied to Pakistan", *Washington Post,* December 21, 2003.

11. Ibid.
12. Ibid.
13. Ibid.
14. Devin T. Hagerty, "The implications of a nuclear-armed Iran in light of South Asia's nuclear experience", in Sumit Ganguly and S. Paul Kapur, eds., *Nuclear Proliferation in South Asia* ( London: Routledge, 2009), p. 216.
15. Gordon Corera, *Shopping for Bombs: Nuclear Proliferation, Global Insecurity and the Rise and Fall of the A. Q. Khan Network* (New Delhi: Foundation Books, 2006), p. 61.
16. Ibid., p. 60.
17. "Nuclear Black Markets: Pakistan, A. Q. Khan and the Rise of Proliferation Networks: A Net Assessment," *IISS Strategic Dossier*, 2007, p. 67.
18. Ibid., p. 67.
19. Dafna Linzer, "Iran Has Offered Nuclear Parts", *Washington Post*, February 27, 2005, at http://www.washingtonpost.com/wp-dyn/articles/A56391-2005Feb26.html
20. Ibid.
21. Douglas Frantz and Catherine Collins, *The Man From Pakistan: The Story of the World's Most Dangerous Nuclear Smuggler* (New York: Twelve, 2007), p. 176.
22. *IISS Strategic Dossier*, n. 17, p. 70.
23. Corera, n. 15, p. 70.
24. Ibid., p. 70.
25. Ibid., p. 71.
26. *Dawn*, February 2, 2004.
27. "A Nuclear North Korea: Intelligence; US Says Pakistan Gave Technology to North Korea", *The New York Times*, October 18, 2002.
28. "Pakistan and North Korea: Dangerous Counter-Trades", *IISS* Strategic Comments, Vol. 8, Issue 9, November 2002.
29. *IISS Strategic Dossier*, n. 17, p. 72.
30. Ibid.
31. David E. Sanger, "Pakistani Says he Saw North Korean Nuclear Devices", *The New York Times,* April 13, 2004, cited in *IISS Strategic Dossier*, n. 17, p. 72.
32. "China Missile Exports and Assistance to Pakistan", *NTI*, at http://www.nti.org/db/china/mpakpos.htm
33. Ibid
34. Ibid.
35. Bill Gertz, "Pakistan Deploys Chinese Missiles", *Washington Times*, June 12, 1996.
36. Sanctions had a significant impact on the sales of high technology goods to China.

Those goods were a major portion of the $12 billion US trade with China in the 1992.

37. n. 28,
38. Corera, n. 15, p. 87.
39. Ibid., p. 87.
40. *IISS* Strategic Comments, n. 28.
41. Ibid.
42. Ibid.
43. "Pakistan's Missile 'Was a No-dong'", *Jane's Missiles &Rockets*, Vol. 2, No. 5, May 1998, pp.1-2.
44. *Jane's Strategic Weapon Systems*, Issue 48, 2008, p. 113
45. See Chapter 4 – Nuclear Delivery Systems.
46. Gaurav Kampani, "Second Tier Proliferation", *The Nonproliferation Review,* Fall/ Winter 2002. Cited in Corera, n.15, p. 90.
47. Corera, n. 15, p. 91.
48. See Sharon A. Squassoni, "Weapons of Mass Destruction: Trade Between North Korea and Pakistan", Updated November 28, 2006, CRS Report For Congress, Order Code RL31900.
49. Larry A. Nilsch, "North Korea's Nuclear Weapon Program", CRS Report, August 1, 2006. Order Code RL 33590.
50. Corera, n.15, p. 92.
51. "North Koreans Went to Pakistan for Nuclear Study: Report", *Reuters,* January 1, 2004.
52. Corera, n. 15, p. 92.
53. Ibid., p. 92.
54. Ibid., pp. 92-93.
55. Pervez Musharraf, *In the Line of Fire: A Memoir* (London: Simon & Schuster, 2006), p. 287.
56. Squassoni, n. 48.
57. Musharraf, n. 55, p. 294.
58. Corera, n. 15, p. 93.
59. Ibid., p. 94.
60. David E. Sanger and William J. Broad, "From Rogue Nuclear Programs, Web of Trails Leads to Pakistan", *The New York Times*, January 4, 2004.
61. Patrick E. Tyler, "Libya's Atom Bid in Early Phases", *The New York Times,* December 30, 2003.
62. Ibid.
63. Sanger and Broad, n. 60.
64. Ibid.

65. Corera, n. 15, p. 107.
66. *IISS Strategic Dossier*, n. 17, p. 76.
67. Corera, n. 15, p. 108.
68. *IISS Strategic Dossier*, n. 17, p. 77.
69. Corera, n. 15, p. 109.
70. Ibid., p. 113.
71. *IISS Strategic Dossier*, n. 17, p. 80.
72. Frantz and Collins, n. 21, p. 301.
73. See Joby Warrick and Peter Slevin, "Libyan Arms Papers are Linked to China", *Washington Post,* February 15, 2004.
74. Frantz and Collins, n. 21, p. 301.
75. Ibid., p. 302.
76. Corera, n. 15, p. 119.
77. Ibid., p. 120.
78. "Pakistan Rules Out Joining Nonproliferation Treaty", *NTI Global Security Newswire,* February 23, 2010, at http://www.nti.org/gsn/article/pakistan-rules-out-joining-nonproliferation-treaty/
79. Ibid.
80. Amit Baruah, "Pakistan's Compulsions", *Frontline,* Vol. 16, Issue 20, September 25-October 8, 1999, at http://www.frontlineonnet.com/fl1620/16200400.htm
81. Rabia Akhtar, "Should Pakistan Sign the CTBT? " September 5, 2009, at http://www.thedailystart.net/newDesign/news-details.php?nid=104394
82. See Naseem Zehra, "Pakistan Security and CTBT", *Media Monitors Network,* June 15, 2001, at http://www.mediamonitors.net/nasimzehra3.html
83. "Letter to the Editor: Pakistan's Conditiona for an FMCT", *Arms Control Association,* January-February 2012, at http://www.armscontrol.org
84. Ibid.
85. *IISS Strategic Dossier*, n. 17, 2007, p. 65.
86. Edidin, n. 2.
87. *IISS Strategic Dossier*, n. 17, p. 67.
88. Corera, n. 15, p. 122.
89. David Blair, "Bleak Future for Pakistan's 'Bomb Hero' ", *Daily Telegraph,* February 7, 2004, cited in Corera, n. 15, p. 122.
90. William J. Broad, David E. Sanger and Raymond Bonner, "A Tale of Nuclear Proliferation: How Pakistan Built His Network", *The New York Times,* February 12, 2004.
91. Sanger and Broad, n. 60.
92. Farzana Shaikh, "Pakistan's Nuclear Bomb: Beyond the Non-Proliferation Regime", *International Affairs*, Vol. 78, No. 1, January 2002, pp. 29-48.

93. "A. Q. Khan (Pakistan) Behind Iraqi Nuclear Programme", October 5, 2002, *NewsInsight.net*.
94. Anwar Syed, "An Affair to Remember & Regret", *Dawn,* February, 15, 2004.

# 3. China-Pakistan Strategic Nexus: Strengthening the Conventional and Nuclear Capability

---

The China-Pakistan relationship stands as one of the most unique and durable relationships between two sovereign states, where both states have been complementing each other's strategic requirements in spite of different cultures, political systems and social values, not to talk of different history. China has emerged as Pakistan's most trusted ally, that has not only supplied Pakistan with a massive amount of conventional military equipment but also provided vital nuclear assistance which Pakistan desired eagerly to neutralise India's conventional military superiority.

The China-Pakistan strategic partnership which started as early as 1951 has continued to grow and both nations have enjoyed the mutually beneficial relationship. There is a growing consensus within Pakistan that its relationship with China is indispensable because of the sustained military, strategic and economic assistance provided by China and also due to the belief that Pakistan and China have common strategic interests. Pakistan's military needs, attributed to its perceived threat from India, have been, by and large, fulfilled by China. Over the years, China has provided Pakistan a wide range of conventional weapon systems, and the two countries have concluded several defence cooperation agreements. While Pakistan turned towards Beijing as a trusted 'all weather friend' in dealing with its 'implacable' enemy, India, China, on the other hand, found a feasible option in Pakistan to contain India and also the expansion of US dominance in the region.

Pakistan, in fact, is one of the first groups of countries that recognised the People's Republic of China (PRC) on May 21, 1951, when the two nations officially established their contacts. Since then, the relationship between the two nations has evolved smoothly and Pakistan has received enormous military, economic and diplomatic assistance from China which has strengthened it militarily, diplomatically and strategically, and certainly boosted its morale.

**The Early Years**

In the early years after the establishment of the Sino-Pakistani diplomatic relations, Pakistan, being a military ally of the West, maintained a general link with China. It was during the Bandung Conference in 1955 that Premier Zhou Enlai held friendly talks with Pakistani Prime Minister Mohammad Ali, and in the exchange of ideas, agreed that cooperation between the two nations should be strengthened in various fields.[1] China's aim was to wean Pakistan away from the military alliance led by the United States against the Soviet-PRC (People's Republic of China) military alliance. The conference assisted in building the relationship between China and Pakistan and initiated a high level exchange of visits between the two nations.[2] Zhou Enlai assured Field Marshal Ayub Khan that there was no need for Pakistan to be apprehensive about China as it had no plans of aggression against Pakistan.[3] Pakistan, on the other hand, was looking for support in order to build its capacity to cope with the Indian military and diplomatic posture. This was very clearly defined by Ayub in his autobiography wherein he articulated his foreign policy:

> China was making significant progress in all spheres of life, and seemed anxious to have friendly relations with her neighbours. All we had to do was to convince her of our sincerity and friendly intent.[4]

He further wrote:

> It was easy to deduce that if we could not establish normal relations with all our three big neighbours, the best thing was to have an understanding with two of them. They might have their internal difference but we did not need to get involved in that……..It was on this basis that I set out to normalise our relations with the People's Republic of China and the Soviet Union. It is in this sense that our geographical locations and the political compulsions inherent therein have determined the course of our foreign policy in recent years.[5]

Pakistan's attitude towards India remained hostile and apart from the differences over ideology and Kashmir, what worried Pakistan was the distribution of water from the Indus river system. The issue of water has been one of the root causes for Pakistan's aggressive posture as early as in 1947 which was defined by Maj Gen Akbar Khan in his writings. Citing the objectives behind Pakistan's first aggression in 1947, Maj Gen Akbar Khan said that Pakistan feared that India might disrupt the water supplies to West Pakistan which would affect the agriculture dependent West Pakistan —Punjab and strangle Pakistan economically.

It would not be incorrect to say that Pakistan's interest in, and approach to, China intensified in 1959 after the Tibetan revolt and the flight of the Dalai Lama to India which embittered the Sino-India relationship. It was as early as 1959 that Ayub signalled to Beijing his interest in demarcating the Sino-Pakistani border in northern Kashmir.[6] A further boost to the Sino-Pak relationship came post Sino-India border war in 1962. China displayed ample interest in extending support to Pakistan despite its membership of the US alliance system against China and the Soviet Union. Pakistan received economic, military and strategic support from China and, in return, China was guaranteed a foothold in the subcontinent and also assurance of support from the Islamic world, west of Pakistan.[7]

The informal Sino-Pak alliance grew stronger and a border agreement was negotiated between the two nations in 1962. In March 1963, the two countries signed the boundary agreement on "China's Sinkiang and the contiguous areas, the defence of which was under the actual control of Pakistan."[8] In the agreement, the two parties agreed to delimit the boundary, "…on the basis of the traditional customary boundary line, including natural features and in a spirit of equality, mutual benefit and friendly cooperation."[9]

In Article 6 of the agreement, interestingly, China made the agreement conditional in that: "the two parties have agreed that after the settlement of the Kashmir dispute between Pakistan and India, the sovereign authority concerned will reopen negotiations with the Government of the People's Republic of China on the boundary……"[10]

The agreement had been under consideration for some time but it certainly received impetus after the Sino-India War in 1962. Khalid Bin Sayeed has talked about the reasons why both China and Pakistan went in for the agreement. Explaining the Chinese stand, he states, "...obvious answer was that since they (PRC) had been branded in the West as an aggressor in their border dispute with India, they were interested in changing this image and presenting themselves as a reasonable and friendly power."[11]

Pakistan was keen to move towards China not only because it needed support against India, but also because of its insecurity vis-à-vis China. The Chinese position on Tibet in 1959 and the subsequent events obviously generated a Communist threat in the region and it would not be incorrect to presume that Pakistan feared Chinese aggression. Khalid Bin Sayeed brought out this aspect in the Pakistani psyche when he wrote:

> One could not say it was merely out of spite for India that such an agreement was arrived at. An important consideration was that Pakistan might be subjected to a similar Chinese attack, particularly when the Chinese had good roads on their side, if a border agreement was not reached. Such Pakistani fears were also disclosed by Foreign Minister Bhutto in his speech in Dacca on April 8, 1963.[12]

Zulfiqar Ali Bhutto also expressed this fear in Pakistan:

> When the Chinese came into Tibet and the Dalai Lama sought refuge in India, relations between India and China began to get strained..........A similar situation could arise on our undemarcated borders in the Sinkiang and Baltistan areas. We had been receiving reports from time to time that Chinese patrols were coming up to Shamshal. There had been no shooting incident, but the Chinese had driven away some cattle in certain areas...[13]

India kept on strongly objecting since 1962 that Pakistan was in illegal occupation of the territory, hence, it would not be incorrect to say that the treaty was illegitimate and that Pakistan sought to give legal status to its illegal occupation. After the Sino-India War in 1962, Pakistan realised that long-term Sino-India hostility would have major repercussions and, thus, it was in its best interest to align with China. On the other hand, the agreement was one part of a coherent Chinese policy to seek support through Pakistan among neighbouring Muslim nations, in view of the unrest among the Muslim tribes in Xinjiang, and the Russian attempts to exploit it. [14]

The agreement becomes very important in the history of the Sino-Pakistan relationship as it provided the base for the future defence and military collaboration between the two nations. Ayub Khan wrote, "This agreement on border demarcation was the first step in the evolution of relations between Pakistan and China.....as a result of this agreement, the Chinese began to have trust in us and we also felt that if one was frank and straightforward, one could do honest business with them."[15]

The growing economic relationship between India and the United States was another factor leading to Pakistan's inclination towards China. The United States continued to express its dissatisfaction over the increasing Sino-Pakistan relations. In July 1963, a Press officer of the US State Department commented on the air agreement between Pakistan and China and said that the air link "could have an adverse effect on efforts to strengthen the security and stability of the subcontinent, which the Chinese Communists want to prevent."[16]

India received nominal or, in fact, no military assistance from the United States. The US had refused to sell arms to India and the only US arms in Indian territory were:

- Weapons and equipment of World War II (mostly vehicles like the jeep).
- Locally resurrected Liberator bombers from the junkyard of aircraft that had crashed earlier.
- Sale of the C119 Packet medium transport aircraft (which were modified locally by fitting a jet engine to provide them the power to fly in/over the Himalayas).
- A few helicopters (like the light Bell-G 3 and S.55 Sikorsky).

It would be interesting to note that the US supplied missile firing F-104 supersonic interceptors and F-86 combat aircraft to Pakistan but, on the other hand, kept rejecting the request for the F-104 sought by India.[17] Even after the 1962 War, India could only receive some winter clothing (for two divisions) and small arms. But Pakistan made a great hue and cry about the US providing "arms" to India in order to justify its new alliance with China and also highlight its perceived insecurity on the world forum.

Pakistan's anxiety and frustration was expressed by the political leadership: "Overlooking the delicate balance of power in the subcontinent, the United States' Government did not wait to consult Pakistan before rendering large-scale military assistance to India in the aftermath of its autumn 1962 border conflict with China."[18]

The 1965 India-Pakistan War marked an important landmark in the Sino-Pakistan relationship, providing it a new dimension. China demonstrated support for Pakistan which was a member of the Southeast Asia Treaty Organisation (SEATO), basically constructed to counter Communist expansion. On the other hand, US-Pakistan ties were undermined and Pakistan faced suspension of military assistance from the West. Although Pakistan faced the US arms embargo after the 1965 War, it was not punished for violating the UN Resolution and launching the war. India raised serious objections to Pakistan using US weaponry in the 1965 War, and complained to Washington that by doing so, Pakistan had breached US assurances to India and also violated the terms of arms transfers under the 1954 Mutual Defence Assistance Agreement.[19] Being a member of the two security alliances, SEATO and Central Treaty Organisation (CENTO), provided Pakistan a stronger claim on the US resources in the late 1950s and early 1960s. By the year 1957, Pakistan was receiving a massive amount of sophisticated military equipment, training and economic aid from the US. The inflow from Washington included sophisticated Patton Main Battle Tanks (MBTs), modern artillery, howitzers, F-86 jet fighter squadrons, F-104 Starfighter supersonic interceptors, air-to-air missiles, submarines,[20] and state-of-the-art radar, communications and transportation equipment. A further qualitative boost came from the military training by US military teams and also in the US military schools to the Pakistan Army.[21] The US military also provided assistance in setting up intelligence and special operations facilities. Although the shift in the US policy towards India and increasing assistance for India was making Pakistan uncomfortable, a major setback for Pakistan was the announcement of stoppage of all US military assistance on April 12, 1967. This obviously created tremendous confusion in Pakistan's defence planning and added to its desperation to look for options in order to acquire the high technology weapons.

## Chinese Military Assistance to Pakistan

China is today Pakistan's largest defence supplier. Pakistan has not only imported the maximum types and numbers of defence equipment from China but has managed to build up its indigenous defence capability with Chinese assistance. Sino-Pakistan military collaboration witnessed significant developments post 1965, which could be attributed mainly to three factors:

First, during the 1965 War, China demonstrated sympathy and support for Pakistan, even though it was a member of SEATO. This created goodwill for China within Pakistan.

Second, while Pakistan failed to win the war in 1965, its military, nevertheless, projected it as a victory, especially in the air, and the thirst for high technology systems intensified.

Third, and most importantly, the US arms embargo led to withdrawal of the US military assistance and also suspension of the supply of defence equipment to Pakistan.

Pakistan turned to China, North Korea, Germany, Italy and France for military aid. In the following decade, although the US equipment was not available to Pakistan, modernisation of the military, especially the Pakistan Air Force, was kept up, and this was mainly due to Chinese sales.

### *Chinese Defence Supplies to Pakistan*

Even as the 1965 War was getting underway, Pakistan sent its recently retired Air Chief, Air Mshl Asghar Khan, to China to seek aircraft and weapon systems to meet Pakistan's "dire needs". Pakistan's need for different sources was complemented by Bhutto's approach which believed in maintaining relations with all the major powers whose interests lay in South and West Asia. Bhutto's diplomatic policy brought Pakistan closer to China and away from the United States. He encouraged Pakistan to enter into several economic and cooperation arrangements with China and the Soviet Union.[22] Although Pakistan finalised deals with France for defence acquisitions, over time, China began to secure a much larger share owing to the following factors:[23]

- Pakistan and China signed three pacts in 1963 which covered trade, civil aviation and borders wherein Pakistan illegally ceded large tracts of land in Shaksgam Valley in Jammu and Kashmir (J&K) to China.
- China demonstrated open support to Pakistan in the 1965 War and also supplied war material, including T-59 tanks and MiG-19 fighter planes, beginning the China-Pakistan military strategic cooperation (even though it was one-sided initially).
- Chinese weapons turned out to be cheaper than the weapons from the West.
- Credit from China was available on easy repayment terms.
- In due course, Pakistan viewed China as a more reliable partner as compared to the US.[24]

***Weapon Supply in the 1960s and 1970s***

Pakistan did manage to acquire high technology equipment from the US in the 1950s and early 1960s which prompted it to go in for its aggression in 1965 against India. It was mainly the US equipment which was used by Pakistan in the 1965 War. In the 1960s, especially the late 1960s and later in the 1970s also, Pakistan received interest free economic aid and also a significant amount of free weapons from China and became the only non-Communist Third World country to receive generous assistance from it.[25] The Chinese F-6 entered the Pakistan Air Force's (PAF's) inventory in 1966 followed by other systems. In the late 1960s, Pakistan received MiG-19 fighters from China, apart from the substantive infantry equipment. China supplied 115 F-6 fighters between 1971 and 1981. Chinese military assistance came in not only in the form of arms but also development of infrastructure for repair and overhaul.

On the naval front, after the 1971 War, the Pakistan Navy opted for a modest acquisition programme in the form of new Chinese built missile/torpedo attack craft. Between 1972 and 1980, 12 Slaughter class attack FPBs, 4 Hainan class attack craft and 4 Huchuan class hydrofoil craft were delivered to Pakistan.[26] Pakistan's naval acquisitions in the late 1970s focussed on building surveillance and targeting capability and, thus, the deal for the French Atlantiques was finalised.

***Supply Since the 1980s***

The Soviet invasion of Afghanistan in December 1979 led to the American review of their South Asian policy, and, consequently, Pakistan entered into a new engagement with the US. Thus, in the 1980s, the American arms pipeline reopened for Pakistan and it received massive aid from the US in return of its status as a "frontline state" for Washington. The US military assistance programme included the sale of 40 F-16 Falcon multi-role combat aircraft, attack helicopters and second-hand destroyers.

Chinese weapons, being cheaper, continued to hold a significant share in the Pakistani inventory (Table 3.1). Although arms from China were technologically not as superior as those from the West, they were capable systems, were affordable, and provided quantity to boost Pakistan's military capability. China (besides France) continued to be a major supplier of PAF weapons. About 90 A-5s were obtained in 1983-84 for the price of $1 million per aircraft. Acquisition of 95 F-7 series aircraft was started, adding to the quantitative element in the PAF.[27] In fact, by the early 1980s, China had provided Pakistan with roughly about 65 per cent of its aircraft and roughly 75 per cent of its tanks.[28]

**Table 3.1: Pakistan's Major Arms Acquisitions in the 1970s and 1980s**

| Supplier | No. ordered/ delivered | Weapon designation | Weapon description | Year(s) of delivery |
|---|---|---|---|---|
| China | 115 | F-6 | Fighter | 1971-81 |
| | 142 | Q-5 Fantan-A | Fighter/grd attack | 1983-86 |
| | 1044 | T-59 | MBT | 1970-86 |
| | 50 | T-60 | Light tank | 1981-82 |
| | 50 & 200 | T-63 & Type 531 respectively | Light tank & APC respectively | 1972-76 |
| | 4 | Hainan Class | PC | 1976-80 |
| | 4 | Hegu Class | FAC | 1981 |
| | 4 | Huangfen Class | FAC | 1984 |
| | 4 | Huchuan Class | Hydrofoil FAC | 1973 |

In the 1980s, the Pakistan Army inventory included significant numbers of Chinese equipment including the T-59 MBTs, T-60 light tanks and T-63 light tank and Type 531 Armoured Personnel Carriers (APC). Although cooperation between the Pakistan Navy and Chinese Navy dates back to the 1970s, there was not much acquisition from China till the late 1980s. In 1984, four Huangfen class missile attack craft were transferred from Beijing for about $20 million per piece. It is interesting to note that the Chinese naval equipment, being inferior in quality, was less desired, but Pakistan obtained the missile craft mainly with a longer-term objective of striking a deal of technology transfer in the future for indigenous production of the missile craft.[29]

**US Sanctions**

American military and economic aid came to a halt following the sanctions in 1990. The sanctions were highly damaging in nature as they not only suspended the US military aid and assistance but the procurement of essential spares was also blocked. Intensive lobbying by Pakistan resulted in some relief under the Brown Amendment, passed in 1995. The amendment permitted Pakistan to take possession of the military equipment frozen in the United States, with the exception of the nuclear capable F-16 combat aircraft.[30] These sanctions actually impacted the PAF's capability and created confusion in the PAF's planning and procurement regarding the replacement of these aircraft.

The decade of the 1990s was a setback for Pakistan's military modernisation due to the American sanctions and also Pakistan's crippled economy. Economic growth recorded a steep decline and Pakistan was under severe pressure from the international financial institutions to reduce the fiscal deficit and, hence, cut down the spending on defence. Despite the US sanctions, low Gross Domestic Product (GDP) and a collapsed democratic structure, Pakistan tried hard to acquire the defence equipment.[31]

Pakistan, with its nationalist ego boosted by the nuclear weapons tests in 1998 (which it also believed would deter India from any robust response), launched the war in the Kargil sector of J&K in early 1999. This misadventure not only led to its defeat but triggered the return of the army in control of the country, displacing the elected government in the coup of October 12, 1999.

With the emergence of the military regime, another set of democracy related sanctions was imposed on Pakistan by the US, laying further restrictions on acquisition of high technology weapons from the West. The result was China's preeminence in Pakistan's arms imports (Table 3.2).

**Table 3.2: Pakistan's Major Arms Acquisitions During 1990-2011**

| Supplier/ Licenser | No. ordered / delivered | Weapon designation | Weapon description | Year(s) of delivery |
|---|---|---|---|---|
| **China** | 250+ | Technology for the licensed production of T-69-II | MBTs | 1991-99 |
| | 80+ | M-11 | Short-range ballistic missiles | 1991-98 |
| | 264, 3 & 1 | T-85II, T-85IIAP & Type 85III | MBTs | 1992-97 |
| | 40 | F-7P | Combat aircraft | 1993 |
| | 6 | K-8 | Trainer aircraft | 1994 |
| | 20 | Armoured combat vehicles | | 1995 |
| | 3 | Type-347G | Fire control radar | 1997-2001 |
| | 27 | 130mm guns | Large calibre artillery systems | 1998 |
| | | | | |
| | 75-100 | K-8 | Fighter/trainer aircraft | 2000 |
| | 60 | 130 mm guns of Type 59-1 | Large calibre artillery systems | 2000 |
| | 2 | Technology transfer for the construction of Jalalat-class FAC | Missile- armed | 2000-2001 |
| | 24 | C-802/CSS-N-8 | Anti-ship missiles | 2000-2001 |
| | 4 | Jiangwei | II class frigates | 2002-2006 |
| | 40 | F-7 MG | Fighter aircraft | 2001 |
| | 44 | F-7 PG aircraft | Combat aircraft | 2001-02 |
| | 3 | Type-347G | Fire control radar | 1997-2001 |

| | | | | |
|---|---|---|---|---|
| | 150+100 =250 | FC-1/JF-17 | Multi-role combat aircraft | China-Pakistan joint venture, delivery possibly started in 2006 |
| | 6 | A-5 | Combat aircraft | 2003 |
| | 143 | 122 mm D-30 (Howitzer) | Large calibre artillery systems | 2003-04 |
| | 2 | Type-347G | Fire control radar | 2004 |
| | 10 | Type 85 | Main battle tanks | 2004 |
| | 2 | Type-347G | Fire control radar | 2004 |
| | 16 | C-802/CSS-N-8 Saccade | Anti-ship missile | 2004 |
| | | PL-12/SD-10 | BVR AAM | Year of order-2004 |
| | 64 | FM-80/HQ-7 | SAM | Year of order – 2005 |
| | | PL-12/SD-10 | BVR AAM | Year of order - 2004 |
| | 4 (delivery 2009-2013) | F-22P type | Frigates | Year of order -2006 |
| | 4-6 | Z 9EC | Helicopters | Year of order-2006 |
| | 4 | ZDK-03 | AWACS | Year of order-2008 |
| | 40 | J-10 | Multi-role Combat Aircraft | Delivery starting 2014-2015 |

*Source:* Based on Stockholm International Peace Research Institute (SIPRI), Yearbook (issues of various years) (New York: Oxford University Press) and Pakistan's major arms imports, as cited in Jasjit Singh, "Trends in Defence Expenditure," *Asian Strategic Review*, 1998-99 (New Delhi: Institute for Defence Studies and Analyses, 1999); *SIPRI 2002, 2004, 2005*; *United Nations Register of Conventional Arms* for various years and Defence Security Cooperation Agency at http://www.dsca.mil; "Update on Pakistan C-130E Acquisitions," *Air Forces*, January 2006, p.22; *The Military Balance 2007, IISS* (London: Routledge, 2007). *The Military Balance 2010, IISS* (London: Routledge, 2010). *The Military Balance 2011, IISS* (London: Routledge, 2011).

Sino-Pakistan defence collaboration flourished under the umbrella of the US sanctions. The Aircraft Manufacturing Factory (AMF) under the Pakistan Aeronautical Complex (PAC) at Kamra, entered into the production of the Karakoram-8 jet trainer, in collaboration with the China National Aero-Technology Import and Export Corporation (CATIC). The jet trainer was initially launched as NAMC L-8 in 1987 in the Paris Air Show and subsequently proposed to be co-developed in partnership with Pakistan (with 25 per cent share). The jet trainer K-8 has been named after the mountain range in Jammu and Kashmir. The upgraded version of the jet trainer K-8P is with an avionics upgrade and has been available since 2005. The PAF received the initial batch of 8 K-8P, followed by further 8 in January 2009. The original 12 K-8 have also been upgraded to K-8P level.

China and Pakistan entered into deals for the co-development of a fourth generation fighter aircraft, the JF-17 (earlier called the FC-1). China delivered 8 JF-17 to the Pakistan Air Force in March 2007 for flight tests and evaluation. The JF-17 is designated to be a low cost, multi-role combat aircraft to meet the tactical and strategic requirements of the Pakistan Air Force, thereby reducing Pakistan's reliance on imports. The JF-17 is co-developed by Pakistan and China and is being built by China's Chengdu Aircraft Industry Corporation (CAC) and Pakistan Aeronautical Complex (PAC), Kamra. There have been reports that the design was developed by the MiG complex in Russia and transferred to China after the Russian Air Force cancelled procurements. Pakistan has also increased its initial target of buying 150 JF-17s to up to 250 aircraft.[32] This represents a quantum jump in the Pakistan aircraft industry.[33]

Interestingly, beginning of May 2011, immediately after Osama bin Laden's killing in Pakistan's high security military compound of Abbottabad, China announced the decision to deliver 50 JF -17s free of cost, to Pakistan, by the end of 2011. The JF-17 is, indeed, a remarkable achievement for Pakistan as it has now entered into the manufacturing of the fourth generation aircraft. Reports suggest that initially only the assembly of the aircraft was being done at Kamra but presently, quite a few parts are being manufactured at Kamra. Pakistan has been marketing the JF-17 quite aggressively and several countries, including Algeria, Sri Lanka, Egypt, and Malaysia have shown keen interest in the acquisition of the aircraft.

In April 2006, the PAF got the approval of the Pakistani leadership to go ahead with the negotiations for the acquisitions of J-10 from China, as part of its fighter force expansion plans. The agreement for the acquisition of 36 J-10s (also called the FC-20s) was finalised in November 2009. The deliveries of the J-10s are expected to be completed by 2013. The Chinese J-10 along with the JF 17, would form the backbone of the PAF.[34]

China has also confirmed the sale of six ship-based medium sized Z 9C helicopters to the Pakistan Navy.[35] In 2008, the deal for the Chinese built airborne early warning aircraft, ZDK-03, was finalised. The Chinese airborne early warning aircraft are scheduled to arrive within four years on a deferred payment basis. Pakistan is the first country to buy the Chinese Airborne Warning and Control System (AWACS) which China started to produce only in 2004. Pakistan has been focussed on the acquisition of force multipliers and this deal with China is in addition to the SAAB turboprop platform equipped with the Erieye, from Sweden. Airborne early warning would certainly be an additional asset for offensive air strikes at sea. The deal for 4 aerial refuellers from Ukraine has been finalised. The Il-78 will give the PAF its first airborne refuelling capability. Once operational, these will be used to refuel the Mirage III and, eventually, the JF-17s. Acquisition of aerial refuelling would provide extended range and deployment radius. The PAF capability would increase significantly by 2020 (Table 3.3).

**Table 3.3: Modernisation Effect: PAF 2020**

| | **1990** | **2020** |
|---|---|---|
| Combat* | **380** | **420** |
| AEW&C | **Nil** | **8** |
| Air-to-Air Refuelling | **Nil** | **4** |
| Maritime Patrol and Air Strike Aircraft | **3** | **18** |

* It is interesting to note that the combat number in the table does not give the true picture. Out of the total combat aircraft in 1990, just around 10 percent were fourth generation. *But by 2020, more than 90 percent would be fourth generation aircraft in the PAF inventory.*

On the naval front, the Chinese provided assistance towards the build-up of indigenous capability. In 1994, a collaborative venture was initiated with the Chinese for the manufacture of gun and missile boats. Construction of the boats was carried out at the Pakistan Navy (PN) Dockyard and Karachi Shipyard and Engineering Works (KSEW); while the technology came from the Chinese, the boats were fitted with components from the West.[36] Pakistan had also started the construction of small craft like coastal tankers, missile boats, Agosta 90-B submarines and minehunters, with the transfer of technology from China and France.

Pakistan's naval acquisitions from China, in the current decade, include 24 C-802/CSS-N-8 anti-ship missiles and 4 Jiangwei II class frigates. In 2006, the Pakistan Navy ordered four F-22P type frigates from China in a deal valued at $600 million.[37] The first destroyer, the PNS *Zulfiqar*, was delivered in 2009, the second destroyer, PNS *Shamsheer* was delivered in early 2010, and the other two are scheduled to be delivered by 2013. The F-22P, which is a modification of a Chinese frigate that uses a Russian-designed main gun rather than a Chinese model, is armed with eight C-802 anti-ship warfare missiles, eight FM-90 Surface-to-Air Missiles (SAMs), one AK-176M main gun and two Chinese 30 mm Close in Weapon Systems (CIWS). The frigates can be loaded with one Z-9EC helicopter.[38] The fourth F-22P will be manufactured in Pakistan at a Karachi shipyard in 2013, to fulfill a pledge to transfer Chinese shipbuilding technology that was part of the April 2005 agreement to build the frigates.

Other defence production plans on the naval front include four modern corvettes which are planned to be built along with the F-22P in KS&EW. The navy also plans to manufacture and procure additional minehunters, tankers, missile and patrol boats.[39]

### China and Pakistan's Arms Industry

Gen Zia-ul-Haq accorded high priority to building up of defence capability. Defence expenditure increased substantially during his tenure and stood at an average rate of 6.5 per cent of the Gross Domestic Product (GDP). He believed in building self-reliance and, thus, several indigenous projects were promoted. Later, the American arms embargo added to Pakistan's quest for self-sustenance in defence equipment. Domestic capability build-up was boosted but this was not at the cost of equipment technology acquisition arrangements with the USA and China, but complementary to them. The nation's defence industry increasingly became the "training post" for technology.[40] The 1980s and 1990s saw a wide expansion of defence production activities and a large number of varied projects were undertaken in this period. China has been the main support in the establishment of defence production units in Pakistan, often provided free of cost.

**Heavy Industries Taxila (HIT)**

This facility is located near Taxila, and was set up with the help of China. Until the mid-1960s, Pakistan was receiving imported tanks, initially from the US (the M-47 and M-48 Patton class in the 1950s and the 1960s) and then from the Chinese (the T-59, which was produced in China from the Soviet T-54 design built under licence).[41] In 1971, a Heavy Rebuild Factory Project (P-711) was conceived with Chinese assistance and technology to rebuild T-59 tanks. As the fleet of T-59 tanks manufactured by China grew, the Heavy Rebuild Factory (HRF) was established in Taxila in the late 1970s, mainly to undertake repair, overhaul and modernisation of tanks.[42] The facility is spread over 48 acres of land and is capable of rebuilding tanks and power packs as well as manufacturing individual optical and electrical components. When the production started in 1981, the HRF had the capacity to rebuild 100 tanks and 250 engines every year, and by October 1990, that capacity had been able to service 1,000 tanks and 2,000 engines.[43] It is claimed that out of 11,000 components used in the overhaul of the T-59, approximately 8,000 are now manufactured locally.[44] More recently, the HRF at Taxila was renamed Heavy Industries Taxila (HIT). The overhauling of the Patton tanks and the upgrading of the T-59s was being done in critical areas, with technical assistance from China, America and the UK in the five key areas:[45]

- The 100mm tank gun which fires standard North Atlantic Treaty Organisation (NATO) ammunition and is manufactured by the Pakistan Ordnance Factory (POF).
- The fighting compartment fitted with a fire/explosion, detection and suppression system which was lacking before.

- Thermal sights fitted in place of the first generation infrared devices.
- A computerised fire control system with a laser range finder.
- A stabilisation system fitted to all stationary and moving targets to be engaged with a high probability of first round hit.

HIT, which employs over 7,000 workers, and is basically devoted to land combat systems, witnessed rapid growth in the 1980s when it started to produce the T-69 MBT with Chinese help. The T-69 11MP tank is fitted with a special engine and also special armour to enhance armour protection.[46] In the beginning of 1989, the then Army Chief, Gen Mirza Aslam Beg made an announcement that Pakistan would come out with its own indigenously built tank within a period of two years. A prototype of the T-69 11 MP was to come out "by the end of December 1990. A protocol was signed with China that included the establishment of the five factories: Hull Fabrication Factory (HFF), Turret Manufacturing Factory, Engine Manufacturing Factory, Gun Barrel Manufacturing Factory and Flexible Machining Centre."[47] HRF's capabilities were substantially enhanced with the acquisition of skills and experience in the field of production facilities and covering firepower, mobility and protection. The current and future planned production of HIT includes various products, some of which are with the Chinese assistance:

- **Tank 'Al-Khalid':** This is the outcome of a decade long Sino-Pakistan effort. In 1988, the announcement for the manufacture of a new MBT (MBT 2000- Al Khalid) with the assistance of China was followed by the agreement signed in 1990 with the China North Industries Corporation (NORINCO).[48] A four-phased approach was adopted by Pakistan instead of building a new MBT.[49]

  The first prototype of the MBT-2000, also referred to as the P-90, was completed in 1991 and subsequently the trials were carried out in 1992. In many respects, the MBT 2000 appears to be almost identical to the Chinese NORINCO Type 90-II MBT. During the development phase of the Al-Khalid, a number of prototypes of the vehicle were built and tested in various configuration. The first 15 pre-production Al Khalid MBTs were handed over to the Pakistan Army in July 2001.[50]

- **Tank 'Al-Zarrar:** This is again a product of Chinese assistance, which came into being to keep the fleet of T-59 and T-69 tanks technologically and operationally compatible with modern tanks. The Al-Zarrar configurations were built to give the user the maximum possible flexibility in making a decision on the systems to be installed in production vehicles.[51] The Al-Zarrar carried the upgrades which not only are an improvement on the combat capabilities of the Type 59 but also an improvement on overall reliability of the vehicle, with reduced maintenance.[52]
- **Infantry Fighting Vehicle'Al-Hamza':** This has been developed with a Chinese-made one-man turret and 25 mm cannon.
- **WZ 653 series ARV:** In 1995, NORINCO (from China) supplied 20 WZ 653 series Armoured Recovery Vehicles (ARVs) to Pakistan. According to reliable sources, Pakistan obtained the licence to manufacture these and the initial batch of 45 vehicles, also referred to as the ARV-W653 in Pakistan, has been completed at HIT.[53]

**The F-6 Rebuild Factory (F-6RF)**

The Pakistan Aeronautical Complex, established in 1973, is dedicated to the overhaul and rebuild of Chinese and French aircraft in the Pakistan Air Force inventory. The F-6 Rebuild Factory i.e. the F-6RF, in the Pakistan Aeronautical Complex (PAC), is an important facility at Kamra, established with Chinese assistance in 1980. The primary purpose of the F-6RF has been the overhaul of the PAF's Shenyang F-6 aircraft and their accessories (Soviet MiG-17, built under licence in China and sold to Pakistan). Apart from the machine tools, this establishment possesses modern technical capabilities for several engineering processes employed in aircraft manufacture such as surface treatment, heat treatment, casting.[54] It also manufactures metallic/rubber parts required during overhaul of aircraft.[55]

**Heavy Mechanical Complex (HMC) Ltd**

The Heavy Mechanical Complex (HMC) Ltd, located in Taxila is not officially part of the defence production establishment, since it is a major heavy engineering subsidiary of the State Engineering Corporation (SEC), controlled by the Ministry of Industries and Production, Government of Pakistan. This was established in 1979 with Chinese assistance and is the largest undertaking of this type in Pakistan.[56] But the Heavy Forge Factory at this complex is significantly important for Pakistan's defence production requirements. HMC is capable of undertaking the designing, engineering and manufacturing of industrial plants and machinery. It also possesses the largest fabrication and machining facilities in the nation with Computer Aided Designing (CAD) capabilities. It manufactures equipment for a large variety of areas including, hydro-electric power plants, thermal power plants, oil and gas processing plants, steel structure and railway equipment, etc.[57]

**Development of Gwadar Port**

Gwadar has been developed with the Chinese assistance and the primary project has been the construction of a deep sea port, expanding its maritime role and to allow trade to and from the landlocked Central Asia. More importantly, the port would have the conversion facilities to allow the movement of natural gas for the Turkmenistan-Afghanistan-Pakistan-India natural gas pipeline when constructed. Gwadar offers the geo-economic and geo-strategic pivot to China and Pakistan. It is strategically located on the southwestern coast of Pakistan amidst three increasingly important regions of the world: South Asia, Central Asia and oil-rich Middle East. Gwadar, overlooking the Gulf of Oman and the entrance to the Persian Gulf region, is just 180 nautical miles from the Strait of Hormuz**.** Thus, Gwadar would eventually emerge as the key shipping hub, providing mass trade to the Central Asian Republics through Pakistan and China, and an important naval base. China's involvement in Gwadar is undoubtedly a response to its emerging energy requirements, China being the world's second largest oil importer. Approximately, 70 percent of Chinese oil supply comes from the Middle East and Africa through the sea. China is expanding its energy procurement efforts and the strategy of a series of ports along the oil shipment routes which would allow China to safeguard and monitor energy flows.

From the military point of view, Gwadar is a decongestion point to the Pakistan Navy and it will provide it a berthing point for its submarines and surface warships. Gwadar port area has been designated as the "sensitive defence zone" by the Government of Pakistan.[58] Although, there has been no official Chinese announcement on the subject, various reports are indicative of Gwadar being a future berthing point for the PLA-Navy fleet of the Indian Ocean, facilitating China's military presence in the region.

Development of a deep sea port has strategic implications for India which were pointed out by the then Indian Naval Chief, Adm Sureesh Mehta in a lecture in Chennai:

1. Being only 180 nautical miles from the exit of Strait of Hormuz, Gwadar being built in Baluchistan coast, would enable Pakistan to take control over the world energy jugular and interdiction of Indian tankers.
2. The pressure for countries to cooperate in the maritime military domain to ensure smooth flow of energy and commerce on the high sea will grow even further.
3. A highway is under construction, joining Gwadar with Karachi, and there are plans to connect the port with the Karakoram Highway, thus, providing China a gateway to the Arabian Sea.[59]

**Chinese Nuclear and Missile Assistance to Pakistan**

One of the most important outcomes of the China-Pakistan strategic nexus is China's extensive support to Pakistan in building up its nuclear capabilities. Nuclear proliferation analyst Gary Milhollin was not wrong when he argued, "If you subtract Chinese help, there wouldn't be a Pakistani program."[60] After the humiliating defeat in the 1971 War and the creation of Bangladesh, the Pakistani leadership started systematic efforts for acquiring nuclear weapon capability. Bhutto propagated the religious ideology to seek assistance for Pakistan's nuclear programme on the global platform. Nuclear weapons were expected to "neutralise" India's conventional military superiority and, hence, increase Pakistan's military and psychological capacity to continue its fight for Kashmir. The nuclear weapon programme started from 1972, and continued to develop in the 1980s under the leadership of Zia-ul-Haq. Pakistan sought external assistance for its nuclear weapon programme and

China has been the most important source for nuclear exports to Pakistan. China allegedly provided direct assistance to Pakistan's nuclear weapon programme in the past, which includes the supply of the warhead designs and also Highly Enriched Uranium (HEU).

China proved to be Pakistan's crucial partner in its nuclear programme: in the 1970s and 1980s there were definitely state-to-state deals and Pakistan's former top Pakistani nuclear scientist Abdul Qadeer Khan, was the key intermediary.[61] Pakistan received help with the reactor, weapon design and nuclear material. A. Q. Khan visited China on a regular basis and the Chinese scientists were present at Kahuta, possibly helping the Pakistani scientist.[62]

In 1986, China also concluded a comprehensive nuclear cooperation agreement with Pakistan and in the same year, reportedly, Chinese scientists began assisting Pakistan with the enrichment of weapon grade uranium. Pakistan also received tritium gas from China, that could be used to achieve fusion in hydrogen bombs and boost the yield of atomic bombs. The amount of gas transferred to Pakistan was apparently sufficient for making 10 nuclear weapons.[63]

China adopted an extremely supportive posture in the development of Pakistan's nuclear weapons and supplied it with a variety of nuclear products and services, ranging from uranium enrichment technology to research and power reactors.[64] China involved the Pakistani scientists in a nuclear test at its Lop Nur test site in 1983. Pakistan's Foreign Minister Yaqub Khan was also reportedly present at the site. The Chashma nuclear power plant project which was previously initiated and cancelled by France in the 1970s, was resumed in the early 1990s. The nuclear plant was under the International Atomic Energy Agency (IAEA) safeguards with the China Nuclear Energy Industry Corporation (CNEIC) as the foreign supplier and the reactor is based on China's first indigenous reactor, the Qinshan-1.[65]

There was significant pressure building up in the region, as well as from Washington, on China, to refrain from providing its nuclear assistance. In 1993, in fact, China and the IAEA signed an agreement to apply IAEA safeguards to the Chinese nuclear power station sold to Pakistan

(INFCIRC/418).[66] Despite these measures and the pressure from the US, Chinese assistance in this direction continued, including the assistance in the construction of a 40 MW reactor at Khushab.[67]

A 1997 report by the Director of the Central Intelligence Agency (CIA) stated that China "was the primary source of nuclear-related equipment and technology to Pakistan" during the second half of 1996.[68] Although, China has always denied its direct assistance in this direction, there are official and unofficial facts and reports which clearly bring out China as the prime contributor in Pakistan's nuclear bomb. Further clarity and conclusive evidence came from the revelations from the Libyan files and very recently, the confessions by A.Q. Khan. He indicated in a letter that China provided Pakistan with nuclear weapon materials and schematics.

"We put up a centrifuge plant at Hanzhong, (China)," Khan wrote in a 2004 letter to his wife. "The Chinese gave us drawings of the nuclear weapons, gave us 50 kg of enriched uranium, gave us 10 tons of UF6 (natural) and 5 tons of UF6 (3 percent)."[69]

China became a full member of the IAEA in 1984, signed the Non-Proliferation Treaty (NPT) in 1992 and Comprehensive Test Ban Treaty (CTBT) in 1996, but continued the assistance to Pakistan. In 1995, the Chinese supplied ring magnets to Pakistan for the production of HEU. Pakistan's economy in the 1990s was on a total downslide with a decreasing GDP growth rate which stood at an average of around 2.5 percent. But, ironically, Pakistan's nuclear programme was kept up and acquisitions from China for the bomb continued.

China plans to build two new civilian reactors near Chashma, adding a new dimension to the troubled history of nuclear exchange between the two nations. China joined the Nuclear Suppliers Group (NSG) in 2004 and under the NSG obligations, it cannot sell nuclear reactors to a non-nuclear-weapon state that does not have full-scope safeguards. Beijing has claimed that it negotiated the deal for the two reactors with Islamabad before it joined the NSG, even though it certainly did not mention the deal when it joined the NSG.

**China's Missile Exports to Pakistan**

China's continued missile technology assistance to Pakistan has contributed largely in Pakistan's ballistic and cruise missile capability build-up.[70] The issue has been a serious proliferation concern for the United States and has remained a contentious one in the US-China relationship. In 1991, China transferred complete M-11 missiles, missile related technology, and manufacturing assistance to Pakistan.[71] The M-11 is a single stage, solid fuelled missile with a reported range of 300 km and payload of 800 kg. China reportedly started discussing the transfer of M-11 missile technology to Pakistan in the late 1980s and the contract for the sale was signed in 1988. This was followed by the US sanctions against China.[72] China did pledge in 1991 and 1994 that it would comply with the Missile Technology Control Regime (MTCR) and would abstain from missile technology transfer. But, in December 1992, reports suggested that China transferred 34 complete M-11 systems to Pakistan. The reports were obviously denied by China.[73] While denying the missile transfer to Pakistan, the Chinese Ambassador to the United States, Zhu Qizhen, on June 27, 1991, openly admitted at the National Press Club in Washington: "We have sold some conventional weapons to Pakistan, including a tiny amount of short range tactical missiles. I think here you call it M-11. We don't call it M-11, but since you say M-11, let's say M-11.....Let's say, if it is M-11, this is within the range of the MTCR."[74]

On July 15, 1991, again, the Chinese Embassy's Press Counsellor, Chen Guoquing, in an open letter to the editors of *The Washington Times*, stated that China has delivered to Pakistan "very small quantities of short-range tactical missiles."[75]

In 1989, Pakistan revealed the development of the Hatf-1, solid propellant unguided rocket and ballistic missile programme. The Hatf-1 is a single stage solid propellant system, with a range of 70 km and payload of 500 kg that could be conventional high explosive, chemical or submunitions.[76] Although Islamabad claims it to be its indigenous production, there are enough reports which provide evidence of the Chinese assistance. Apart from China, certain European companies also provided assistance to Pakistan.[77] There have been reports on the developments of improved versions, the Hatf-1 A and Hatf-1 B, with Chinese assistance. The Hatf-2, which is also termed Abdali, is reportedly an upgraded version of the Hatf-1B and was developed with Chinese aid and technical assistance. In May 2002, Pakistan made the first flight test for the Hatf-3 (Ghaznavi) missile. Technical evaluation of the missile images suggest that the Hatf-3 is a version of the M-11 or may be even a repainted M-11. Reportedly, the production facilities of these missiles have been set up with Chinese assistance.[78]

Similarly, the Hatf-4 (Shaheen-1), last tested in 2006, is believed to be a scaled up version of the M-11s. The Shaheen-1 is a single stage, solid propellant missile with an inertial guided system and has a maximum range of 750 km.[79] In March 2004, Pakistan test-fired the nuclear capable Shaheen-2 or Hatf-6 ballistic missile, which is believed to be based upon the earlier Chinese two-stage solid propellant missile, the M-18. The maximum range of the Shaheen-2 missile was 2,000 km, which has now been increased to 2,500 km, sufficient to target any important part of India.[80]

Pakistan conducted test-firing of its latest domestically manufactured (as it claims) cruise missile, known as the Babur (or Hatf-7), in May 2009, with a range of 500 km.[81] This test has been the latest in the series of flight tests which commenced in August 2005, followed in 2006 and later in 2007. The development of the Babur, which is an air, ground, ship and submarine launched short range, turbojet powered, single warhead cruise missile, commenced reportedly as early as 1990. In July and August 1998, two US RGM/UGM-109 Tomahawk cruise missiles were recovered almost intact in southern Pakistan, and it is believed that these may have been used for reverse engineering or even to contribute basic technology for the development of the Babur.[82] The missile appears similar to the Chinese Hong Niao-3, the US RGM-109 and also has similarities with the Russian SS-N-27 Club (3M14 version) cruise missiles.[83]

Pakistan's air launched cruise missile Ra'ad (Hatf -8) was tested in May 2008. The origin of Ra'ad is not confirm and the missile, with a range of 350 km, has been developed exclusively for launch from aerial platforms. Reports suggest that in the future, the Ra'ad would be integrated with, and launched from, the JF-17s and the J-10s.

Chinese assistance in missile transfer technology to Pakistan accelerated after Pakistan's nuclear tests in 1998. Pakistan's Surface-to-Air Missile (SAM)

Anza is also reportedly the Pakistani version of the PRC supplied SAM.[84] Other assistance from China includes supply of ammonium perchlorate—a chemical used in rocket fuel—in the mid-1990s,[85] and special metal and electronics used in the production of Chinese design anti-tank missiles.

**Support to Pakistan in its Dispute with India and Wars Against India**

China's position on Kashmir underwent many twists and turns till October 2009, when China took the decision of issuing separate visas to Indian passport holders from J&K and also, very importantly, projecting the disputed territory as an independent country in other ways. Visitors to Tibet—journalists in particular—have received handouts wherein Kashmir is indicated as a country separate from India.[86]

China has been shifting its stand on Kashmir in favour of Pakistan. The 1963 Sino-Pak Frontier Agreement allowed transfer of Shaksgam Valley of J&K to China, which was certainly legitimisation of Pakistan's occupation of parts of J&K (south of Xinjiang). However, accepting (in 1962-63) that J&K was a disputed area and Pakistan did not have sovereign rights over the Valley, China made the treaty conditional to the final decision on sovereignty of the area south of Shaksgam. The Chinese stand on Kashmir, which turned in favour of Pakistan in early 1964, was certainly one of the motivating factors for the Pakistani thinking which believed in exploring a military solution for Kashmir. Air Cmde Jasjit Singh has very rightly stated China's stand on Kashmir: "In February 1964, China had changed its two-decade-old stand and now sided with Pakistan's position of a solution through a plebiscite in Kashmir, but without specifying the prior withdrawal of Pakistani forces from the state, as required by the UN Resolutions."[87]

In 1964, China accepted Pakistan Occupied Kashmir (PoK) as part of Pakistan and later constructed the Karakoram Highway. China extended support to Pakistan during and after the 1965 and 1971 Wars. During the 1965 War, China not only supported Pakistan diplomatically but also provided equipment for the fighting. Asghar Khan went to Beijing in the middle of the war, requesting China for military help. On September 19, during Ayub's and Bhutto's secret meeting in Beijing with Chou Enlai, accompanied by Marshal Chen Yi, the Chinese Defence Minister, Chou had not only "agreed to his request to up the ante"...but also "recommended that Pakistan continue fighting even if they had to retreat to the hills (like the Communist Party had done during the Long March)."[88]

Chinese weapon supply to Pakistan obviously grew at a faster pace after the 1965 War, facilitating Pakistan with relaxed payment modes for the acquisitions. During the 1971 War, Pakistan did not receive any military equipment from China, but it did provide Pakistan economic, political and moral support after the 1971 War. In 1972, China used its first ever veto to hold up the recognition of Bangladesh as a gesture of support to Pakistan.

In the 1980s, China's position turned less aggressive in its support to Pakistan on Kashmir and it held the view that Kashmir was a bilateral issue between India and Pakistan that should be solved peacefully. But during the 1980s, China's nuclear assistance to Pakistan strengthened Pakistan's aggressive posture and the Pakistan Army Chief Aslam Beg declared the doctrine of "offensive defence" in 1989. With the acquisition of the nuclear bomb, which clearly was due to Chinese assistance, Pakistan's covert activities increased tremendously not only in the Kashmir Valley, but also in other parts of India (terrorism in Punjab). Even though China adopted a neutral posture on Kashmir in the 1980s, its nuclear assistance did add to Pakistan's illegal activities in Kashmir.

During the Kargil War, China maintained absolute neutrality and Pakistan did not receive any direct military assistance during the war. Post Kargil War, Pakistan has been extremely focussed on building up its air force, which it realised was one of the reasons for its defeat, and the maritime strike capabilities of its navy, primarily with the Chinese supplies and the US equipment post 9/11. The Chinese JF-17 and the J-10 would presumably form the backbone of the Pakistan Air Force.

The Chinese stand on Kashmir is now witnessing a significant change which has obviously added to New Delhi's concerns and discomfort.[89] In August 2009, China signed a Memorandum of Understanding (MoU) with Pakistan on building a 7,000 MW hydropower project in Bunji in PoK. Also, China has agreed to build the Diamir-Bhasha Dam on the Indus river in PoK.[90] It is reportedly engaged in more than 100 projects in PoK and has employed around 10,000 soldiers in the region.

Apart from mounting pressure on India over Arunachal Pradesh, China's latest disturbing act of issuing separate visas to Kashmiris clearly projects China's posture of treating/questioning/protesting Kashmir's status as an integral part of India. The recent moves by China have added to Pakistan's psychological and strategic gains on the Kashmir front.

**Implications for India**

China continues to be perceived as the most reliable ally of Pakistan. Pakistan's military modernisation and its evolving military muscle owes significantly to the Chinese assistance. China supported Pakistan during and after the India-Pakistan Wars in 1965 and later in 1971. Pakistan's association with China started in the mid-1950s and was intensified in the 1960s. Pakistan's reliance on China has remained unaffected with the multiple change of regimes in Pakistan, and both military and civilian leaderships have been equally enthusiastic about Chinese assistance in various fields.

The Sino-Pak nexus and Chinese assistance to Pakistan in varied fields have multiple implications for India. For the last six decades, Pakistan has been following the strategy of covert war against India. The reason why Pakistan adopted the route of covert war has been India's conventional military superiority which has deterred Pakistan in its illegitimate claim for J&K. Even with the realisation of India's military strength, Pakistan, in its offensive postures, launched three wars against India and lost all of them. The prime reason why Pakistan opted for the Chinese assistance was the fact that it needed to build up its conventional force and with the suspended Western supplies, China was the only option. The study of Pakistan's covert activities in the Indian territory very clearly brings out the fact that these increased with the build-up of Pakistan's conventional military capability. China has contributed significantly in building up Pakistan's conventional military capability in the previous three decades, which undoubtedly implies that it has enhanced, and will further enhance, not only Pakistan's capability, but also the will to carry on with covert war through terrorism in India, without the fear of being defeated in a retaliatory Indian aggression.

Although, Pakistan has been rewarded heavily (for the third time since 1954) by the US in the last ten years and has received substantive US equipment, it no longer feels threatened by any future US arms suspension. Chinese equipment, which now carries Russian technology also, is not inferior to the weapons from the West, and, thus, is adding not only quantity but also quality to the Pakistan's inventory.

China has been active in providing training to the Pakistani officers. Reportedly, thirteen Chinese military institutes are involved in the training of Pakistani officers. These include: College of Defence Studies (NDU Beijing), Army Command College (Nanjing), Artillery College (Nanjing) Military Economics Academy (Wuhan) and Logistical Engineering University (Chongging).[91]

Chinese nuclear assistance to Pakistan has been a matter of deep concern and has altered the security situation globally. Pakistan's nuclear posture, has generated deep concern, with the risks and dangers of a nuclear holocaust. Chinese assistance in building up Pakistan's long-range nuclear missile capability poses a political and psychological threat to India. Even though the possibility of a nuclear war is a remote phenomenon, Pakistan's acquisition of nuclear weapons has strengthened its confidence to continue its strategy to "bleed India".

China's assistance in the development of the Gwadar port is an issue of international attention and mostly interpreted in terms of China's strategy of the "String of Pearls" in areas far from its territory. But what is perhaps of greater significance for India is the implication of a much expanded and technologically advanced Pakistan Navy (and its aerial capability) deployment at the port abeam the major trade and oil transportation routes, besides its proximity to the Strait of Hormuz through which passes more than 50 per cent of the world's crude oil. The bulk of India's oil supplies coming from the Persian Gulf through this choke point would then pose a new challenge of vulnerability that would have to be addressed on priority. Similarly, more than four million Indian expatriates work in the Arab states of the Gulf region. The political implications of potential influence/control by Pakistan on the sea routes between India and the Gulf ports would have to be carefully examined.

Pakistan, with the full knowledge of the reality that India had superior military capability, launched a number of covert and overt wars against India since 1947. China played a crucial role in holding out threats to India during the 1965 War (including deploying military forces on the border and opening fire at Indian troops) launched by Pakistan on September 1, 1965, after three weeks of covert war in Kashmir. This had forced India to: (i) deploy and hold back a substantive portion of its military power for the Eastern Theatre (the Indian Air Force deployed 13 out of 28 combat squadrons in the east). Thus, the full weight of the Indian military could not be applied and a ceasefire was accepted before the Pakistani military could be totally destroyed; (ii) delay the response to Pakistan's genocide till the winter (when Pakistan attacked), and enter into a treaty with the USSR for security consultations.

While China remained neutral during the Kargil War in 1999, no doubt because of the unquestionable transgression and aggression by Pakistan, there is no reason to assume that China will stay neutral in any future conflict. This poses a serious challenge to Indian defence planners: should they plan for a two-front war or not? The mere threat of such a war would necessitate very heavy investments in defence, in which case, there would be a slowdown in economic growth, thereby slowing down India's rise to power.

** A brief version of this paper was presented at the ORF seminar on China on April 30, 2010.*

**Notes**

1. "The Bandung Conference of 1955", *CHINAdaily,* at, http://www.chinadaily.com.cn/english/doc/2005-04/20/content_435929.htm
2. In October 1956, at the invitation of the Chinese government, the then Pakistani Prime Minister H. E. Suhrawardy visited China and in December 1956, Chou Enlai visited Pakistan.
3. Although China had been showing certain parts of Pakistan as part of China in the 300-mile-long undefined border between Xinjiang and Pakistan Occupied Kashmir.
4. Mohammad Ayub Khan, *Friends Not Masters: A Political Autobiography* (London: Oxford University Press, 1967), p. 118.
5. Ibid.

6. Dennis Kux, *The United States and Pakistan 1947-2000: Disenchanted Allies* (New York: Oxford University Press, 2001), p.113.
7. Samina Yasmeen, "China and Pakistan in a Changing World", in K. Santhanam and Srikanth Kondapalli, eds., *Asian Security Review 2003, Asian Security and China 2000-2010* (New Delhi: Shipra Publications, 2004), p. 310.
8. *The Boundary Agreement Between China and Pakistan, 1963.*
9. Ibid.
10. Ibid.
11. Khalid Bin Sayeed, "Pakistan's Foreign Policy: An Analysis of Pakistani Fears and Interest" in Hameed A.K. Rai, ed., *Readings in Pakistan's Foreign Policy*, Vol.-1 (Lahore: Aziz Publishers, 1980), p. 55, as cited by Satyabrat Sinha, "China in Pakistan's Security Perceptions," in Swaran Singh, ed., *China-Pakistan Strategic Cooperation* (New Delhi: Manohar, 2007), p. 82.
12. Ibid.
13. Khan, n.4, p. 161.
14. Sinha, n.11, p. 82.
15. Khan, n.4, p. 164.
16. Zulfiqar Ali Bhutto, *The Myth of Independence* (London: Oxford University Press, 1969), p. 65.
17. Jasjit Singh, *The ICON: Marshal of the Indian Air Force Arjan Singh, DFC* (New Delhi: Knowledge World, 2009), p. 162.
18. Ibid., p. 68.
19. Ibid., p.149.
20. The first submarine to be introduced into the Indian Ocean by a developing country, was the F-104.
21. Dennis Kux, *The United States and Pakistan, 1947-2000* (Oxford: Oxford University Press, 2001), pp. 86-87.
22. Shahid Javed Burki, *Fifty Years of Nationhood* (Lahore: Vanguard, 1999), p. 202.
23. Pakistan's weapon systems sourced from China now account for nearly 75 percent of the total inventory.
24. Pervez Iqbal Cheema, *The Armed Forces of Pakistan* (Oxford: Oxford University Press, 2002), p.164.
25. Ayesha Siddiqa-Agha, *Pakistan's Arms Procurement and Military Buildup, 1979-99: In Search of a Policy* (London: Palgrave, 2001), p. 105.
26. "Pakistan Navy" at http://www.globalsecurity.org/military/world/pakistan/navy-intro.htm
27. Siddiqa-Agha, n. 25, p. 161.
28. Cheema, n. 24, p. 164.
29. Siddiqa-Agha, n. 25, p. 159.

30. For details, see Kux, n. 21. Pakistan had paid for 28 F-16s, which were manufactured against the 1987 order of 110. But following the sanctions, the F-16s were not supplied.
31. In 1990, 50 Mirage 3 (as indicated in Table 3.2) were acquired from Australia for a paltry sum of $28 million, along with engines and spares.
32. Interview, Air Chief Marshal Tanvir Mahmood Ahmed, Pakistan's Chief of Air Staff, April 4, 2007, *Jane's Defence Weekly*, April 4, 2007, p. 34.
33. The PAF inventory currently possesses 16 JF-17s.
34. n. 32.
35. *The Military Balance 2005-2006, IISS* (London: Taylor and Francis, 2005), p. 230.
36. Siddiqa-Agha, n.25, p.165.
37. Bilal H. Khan, "Pakistan Navy Modernization Program", Pakistan Military Consortium, at http://www.pakdef.info/pakmilitary/navy/pn_,odernization.html
38. Ibid.
39. "Pakistan Navy" *Pakistan Defence*, at, http://www.defence.pk/pakistan-navy/
40. Aabha Dixit, "Defence Production in Pakistan," *Asian Strategic Review* 1991-92 (New Delhi: IDSA, 1992), p. 287.
41. John Kaniyalil, " Defence Industries in Pakistan", *Strategic Analysis*, May 1993, pp. 229-247.
42. "FAS Taxila", http://www.fas.org/nuke/guide/pakistan,facility/taxila.htm
43. Mushahid Hussain, "Pakistan's Heavy Armour; Upgradation/Modernisation Programme," *Defence Journal*, October 1990, p. 7, as cited by Dixit, n. 40, p. 293.
44. Kaniyalil, n. 41, pp. 229-247.
45. *Muslim* (Islamabad), July 25, 1991 as cited in Kaniyalil, n. 41.
46. Dixit, n. 40, p. 293-94.
47. Maj Gen Farhat Ali Burki, "A Step Towards Self-Reliance," *Defence Journal*, October 1990, p. 28, as cited by Dixit, n. 40, p. 294.
48. *Jane's Armour and Artillery 2005-2006*, "MBT 2000 (Al Khalid)", p. 74.
49. Ibid.
50. Ibid.
51. The first batch of 80 indigenously modernised and upgraded Al-Zarrar tanks were handed over to the Pakistan Army at the Heavy Industries Taxila in February, 2004.
52. *Jane's Armour and Artillery 2005-2006*, p. 77.
53. Ibid., p. 77.
54. "Indigenisation of Pakistan Air Force's Weapon Systems", at Internet site, http://www.geocities.com/CapitolHill/3023/defence_production.html
55. Ibid.
56. "FAS Taxila", http://www.fas.org/nuke/guide/pakistan,facility/taxila.htm

57. Ibid.
58. "PNS Gwadar," at http://www.globalsecurity.org/military/world/pakistan/pns-gwadar.htm
59. http://intelligencebriefs.blogspot.com/2008/01/
60. Gordon Corera, interview with Gary Milhollin, Washington, D.C., December 14, 2005, as cited in Gordon Corera, *Shopping for Bombs* (New Delhi: Foundation Books, 2006), p. 45.
61. Ibid., p. 45.
62. Ibid.
63. "Incidental Intelligence: China Emerging as Third World Weapons Supplier," *Asia-Pacific Defence Reporter*, September 1991, p. 19; Gary Milhollin and Gerard White, "Bombs From Beijing: A Report On China's Nuclear And Missile Exports," May 1991, p. 17; *The Risk Report*, May 1995, p. 8; Gary Milhollin and Gerard White, "A New China Syndrome: Beijing's Atomic Bazaar," *Washington Post*, May 12, 1991, pp. C1, C4, as cited in, "China's Nuclear Exports and Assistance to Pakistan", *NTI*, at http://www.nti.org/db/china/npakpos.htm
64. See, Chapter 1- Pakistan's Nuclear Programme: The Build-up.
65. "Pakistan's Nuclear Weapons Programme Development", at http://nuclearweaponarchive.org/Pakistan/PakDevelop.html
66. "China's Nuclear Exports and Assistance to Pakistan", *NTI*, at http://www.nti.org/db/china/npakpos.htm
67. Ibid.
68. Ibid.
69. "Pakistan Received Chinese Nuclear Weapon Assistance, Khan Letter Asserts", *NTI*, at http://www.globalsecuritynewswire.org/siteservices/print_friendly.php?
70. For details on Pakistan ballistic and cruise missiles, see chapter 4 – Nuclear Delivery Systems.
71. "China Missile Exports and Assistance to Pakistan", *NTI*, at http://www.nti.org/db/china/mpakpos.htm
72. Ibid.
73. Ibid.
74. *Reuters*, June 27, 1991, as cited in John Wilson Lewis and Hua Di, "Chinese Ballistic Missile Programs: Technologies, Strategies, Goals," *International Security,* Vol. 17, No. 2, Fall 1992, p. 37, as cited in Ruchita Beri, "Pakistan's Missile Programme" in Jasjit Singh, ed., *Nuclear India* (New Delhi: Knowledge World, 1998), p. 198.
75. *Xinhua Domestic Service*, July 16, 1991, see also *JPRS Proliferation Issues*, August 8, 1991, p.2, as cited in Beri, Ibid., p.198.
76. Ibid.
77. *Jane's Strategic Weapon System,* Issue 48, 2008, p. 108.

78. Ibid.
79. Ibid., p.112.
80. Ibid., p. 115.
81. "Pakistan Test-fires Babur Cruise Missile," at http://forum.pakistanidefence.com/lofiversion/index.php/t82549.html
82. *Jane's Strategic Weapons System*, August 3, 2009, p. 117.
83. Ibid.
84. Mushahid Hussain, "Pakistan Missiles; First Pictures," *Jane's Defence Weekly,* March 11, 1989, p. 386.
85. *Jane's Intelligence Review*, October 2, 1996, p. 16.
86. "China Opens a New Front in Kashmir", *Asia Times South Asia,* October 21, 2009
87. Singh, n. 17, p. 167.
88. Ibid., p. 229.
89. "India Warns China Over Kashmir", *asiaone,* October 14, 2009, at www.asiaone.com.sg
90. Ibid.
91. Discussion with Jayadev Ranade, Distinguished Fellow at the Centre for Air Power Studies, New Delhi, on January 12, 2012.

# 4. Nuclear Delivery Systems

Pakistan's efforts to build the delivery systems for its nuclear weapons took off as early as in the 1970s when the Pakistani leadership seriously initiated its efforts to build the bomb. Pakistan's present nuclear dyad consists of nuclear capable combat aircraft and solid fuel and liquid fuel rockets for short range and long range ballistic missiles. Efforts for the acquisition of the combat aircraft like the F-16s (after the failed attempts to get the A-7s in the 1970s) materialised to some extent in the early 1980s, although Pakistan was unable to receive the desired number of F-16s following the US sanctions in 1990. However, Pakistan has focussed on building its nuclear delivery capabilities largely with the Chinese assistance since the 1980s.

Pakistan began its missile programme in the early 1980s as part of its nuclear programme which started in the 1960s. The Pakistani missile programme saw a major upswing in the 1990s that can well be termed as the missile acquisition decade, as in a very short period of time, the country was able to acquire long range potent missiles with nuclear delivery capabilities. Pakistan's missile programme has come into being totally with external assistance, mainly from China and, to a large extent, from North Korea.

Pakistan's nuclear delivery systems can be put under three categories:

- Land-based missiles.
- Aircraft.
- Cruise missiles.

## Land-Based Missiles

### *Pakistan's Missile Development*

Pakistan's deep interest and involvement in missile development can be traced back to as early as the 1960s when the Shah of Iran opened the Iran Electronics Industries (IEI) missile repair facility in Shiraz to Pakistan.[1] Pakistan's efforts towards launching the space programme also date to 1961 when the Space and Upper Atmosphere Research Commission (SUPARCO) was formed. Although Pakistan's efforts to acquire nuclear weapons started in earnest in the early 1970s after Bhutto's announcement to build a bomb, its serious search and efforts for the missiles as a launch vehicle took place in the late 1980s when its nuclear weapon programme was nearly under completion.

Ballistic missiles in a loose sense are more advanced than manned aircraft and are easier to employ, at a relatively lower cost. Pakistan's adoption of an aggressive missile acquisition programme can be attributed to mainly the following factors.

First, Pakistan has always felt threatened by India's superior military build-up and has sought ways to attain parity with India's conventional military capability. In fact, Pakistan's openly stated objective behind going nuclear has been to neutralise India's defence capability. Pakistan tried to acquire the F-16s in the 1980s itself when it was getting close to the completion of its nuclear weapon programme. But after the delivery of 40 F-16s, the US sanctions designed to control Pakistan's growing nuclear weapon programme in the 1990s suspended the delivery of the remaining F-16s. Thus, the Pakistani leadership started to look into the alternatives and missiles came up as an easy option, especially with the growing military partnership it shared with China. Support from China has been at an accelerated pace after the 1965 India-Pakistan War and imports from China have gone up significantly, starting in the early 1990s.

Secondly, while it tried hard to acquire the Mirage-2000 from France, Pakistan's economic conditions in the 1990s did not allow the acquisition of modern combat aircraft from any other source after the US sanctions were levied on Pakistan. Pakistan's GDP growth in the 1990s went down to as low as 1.9 percent (in 1996). Also, there was severe pressure from the International Monetary Fund (IMF) to cut down defence expenditure. There was hardly any acquisition in the 1990s and the PAF actually suffered from lack of modernisation. In the 1990s, Islamabad saw missiles as a cheaper and more reliable option.

Thirdly, Pakistan acquired nuclear capability in 1987 and already possessed F-16s to be used as the delivery vehicles for nuclear warheads. But the aircraft delivery systems have limitations in terms of both range of delivery of nuclear weapons and also penetration of India's air defence system. Ballistic missiles provide a much more credible deterrence, against which there is no credible defence. It is the vulnerability of a nation to a missile attack that lies at the heart of deterrence. This is why Pakistan has focussed on building up the longer range missiles. Missiles are increasingly being touted as the new currency of power.[2]

Lastly, during the Afghan War in the 1980s, Pakistan was exposed to the threat of conventionally armed short range missiles from the Soviet Union. The Soviet forces fired Scud missiles across the Durand Line targeting the Mujahideen camps inside the Pakistani territory. Added to this was the demonstrative threat of the missiles during the Iran-Iraq War in the 1980s. Thus, the evolving role of the missiles was sufficient to influence the Pakistani thinking for a missile development programme.

In the development of its missile programme, Pakistan has clearly followed a dual-strategy:

- Assembly of imported missiles in Pakistan from both the Semi-Knocked Down (SKD) kits and Completely Knocked Down (CKD) kits.[3] This would enable a quantitative jump in the Pakistani indigenous missile production expertise.
- Indigenous fabrication of the above missiles sub-systems and propellants in a graduated manner. This was aimed to create self-reliance in missile capability over time.[4]

***Ballistic Missiles***

Pakistan's missile development programme has been primarily carried out with China's assistance and, to some extent, North Korea's, after the United States imposed sanctions on China. China had been willing to transfer nuclear weapons to Pakistan as early as 1965 itself, shortly after it became a nuclear power.[5] Bhutto had been arguing for the nuclear weapons in the late 1960s, but Ayub had turned down the proposal as it was not deemed necessary or affordable at that point of time for Pakistan. Since 1972, Pakistan sought Chinese assistance for its nuclear development programme and China has been Pakistan's predominant source of foreign technological support for its missile development efforts. Chinese missile assistance to

Pakistan ranges from providing equipment and training to transferring the complete missiles. In 1989, Lt Gen Talat Masood, Chairman of the Pakistan Ordnance Factories (POF) Board (1981-88) acknowledged China's active role in Pakistan's missile programme in an interview.[6] US Assistant Secretary of State Winston Lord was not wrong when he wrote a letter to Senator Robert F. Bennett stating, "The entire strategic weapons program should be stamped 'Made in China'".[7]

***Transfer of M-9 and M -11***

The development of the Chinese M series of Short Range Ballistic Missiles (SRBMs) commenced in the early 1980s and the three versions are known as the M-9, M-11 and M-18. These designations were apparently used for the export versions.[8] All M series missiles use solid fuel, and have short operational preparation time. Also, they are transported by highly mobile cross-country transporters which have the capacity to launch the missiles.[9] Information from various sources indicates that Pakistan had negotiated the deal for the M-11 during Zia's regime. According to US officials, Pakistan agreed to pay China $15 million as partial payment on its 1988 contract with the state-owned China Precision Machinery Import-Export Corporation (CPMIEC) for an undetermined number of M-11 missiles, launchers, and support equipment.[10] China reportedly started discussing the transfer of M-11 missiles to Pakistan in the early 1990s. In the same period, the National Development Complex (NDC), a subsidiary of the Pakistan Atomic Energy Commission (PAEC), reportedly acquired complete though unassembled M-11s and possibly an undisclosed number of M-9 SRBMs from Beijing.[11]

Although Pakistani leaders have made conflicting statements regarding the receipt of the missiles, various reports provide enough evidence that Pakistan did receive the M-9 and M-11 missiles from China. The M-9 SRBMs, which are capable of carrying both nuclear and conventional warheads, were delivered to Pakistan in early 1991. The M-9 is reported to have a range of 600 km. It is a single stage missile with an inertial guidance system, which signifies that the missile is programmed before the launch and does not receive any external guidance after the launch. China has never openly admitted selling the M-9 missiles to Pakistan, but has indicated that it would sell missiles to "whoever can pay for them".[12]

Pakistan reportedly received the M-11 missile (which the Chinese refer to as the Dong Feng-11) in 1991, when US intelligence discovered their transfer along with the accompanying transporter-erector-launchers to Pakistan.[13] The M-11 is capable of carrying nuclear as well as conventional warheads. Pakistan received the M-11 variant as a single-stage, solid fuelled missile with a range of 300 km, carrying an 800 kg warhead.[14] Discussions on the possible sale of M-11 missiles and related technology to Pakistan started in the late 1980s and apparently the contract was signed in 1988. The US intelligence agencies reported in 1995 that the M-11 deal moved ahead after Pakistan apparently paid $15 million to China for the missiles, launchers and support equipment. The M-11s were shipped to Pakistan in 1993, but their assembly was not confirmed.[15] The US spy satellite photographs taken in April 1995 showed missile canisters at a facility in Sargodha, Pakistan. China also sent two teams of missile technicians to Pakistan to provide training, unpack and assemble the M-11s.[16]

The United States' response has not been very consistent in its actions regarding proliferation of nuclear weapons from China to Pakistan and has been driven by its own strategic interests from time to time. The US Administration took measures to dissuade China from transferring the missiles, components and missile related technology to Pakistan, and was engaged in a series of negotiations seeking to obtain pledges from China to abide by the Missile Technology Control Regime (MTCR) guidelines (formulated in 1987) and also to become a member of the regime.[17] Washington offered economic incentives, took diplomatic measures and also imposed economic sanctions on China. Beijing made pledges in 1991 and 1994 when it signed the 31-nation MTCR as well as a 1994 US-China agreement not to deploy M-11s in Pakistan.[18] After Washington received the reports of the transfer of the M-11 to Pakistan, limited sanctions were imposed on China in 1993.[19] The sanctions, affecting an estimated $500 million in American sales, were lifted in October 1994 after Chinese Foreign Minister Qian Qichen and Secretary of State Warren Christopher signed an agreement halting the sale of the M-11 and similar missiles.[20] Various reports at the international level have claimed that China sold over 30 M-11 missiles to Pakistan, despite repeated denials by both governments. On May 8, 1993, the Chinese Foreign Ministry denied the news reports of sale of missile components to Pakistan and announced that the "news report on China shipping M-11 missiles is groundless."[21] At the same time, on May 21, 1993, Secretary General of Pakistan's Foreign Ministry Akram Zaki, denied that Pakistan had acquired M-11 missiles from China. He also termed the news reports of the transfer of the M-11 to Pakistan as "speculative stories and motivated allegations".[22] But, later in the same year, Pakistani Prime Minister Benazir Bhutto stated in Beijing that Pakistan has purchased Chinese M-11 missiles due to the threat posed by the Afghan Scuds and the Indian missile build-up.[23] Again, in 1996-97, there were reports of Chinese assistance in indigenous Pakistani M-11 production. [24]

The M-11 is a road-mobile, single-stage, solid propellant, short-range ballistic missile. The basic variant of the M-11 termed Dong Feng in China has a range of 280-350 km and delivers a single warhead of 500 kg. The basic variant of the M-11 uses an inertial guidance plus terminal radar guidance, giving a Circular Error Probability (CEP) of 500-600 m. The improved M-11A (which Pakistan is reportedly producing) uses inertial/Global Positioning System (GPS) guidance system with optical correlation terminal targeting, resulting in a greater accuracy of below 200 m CEP.[25]

Pakistan has a missile factory located in Rawalpindi for the manufacture of medium range ballistic missiles. In all likelihood, it is manufacturing the M-11 or a similar missile, it is unclear whether this facility has the capability to manufacture complete missiles or only some major components. However, there have been confirmed reports of the continued Chinese assistance during 1996-97, including the blueprints and the construction equipment.[26]

### *Hatf-1*

Development of the Hatf-1 solid- propellant unguided rocket and ballistic missile programme started in the early 1980s and was revealed by Pakistan in early 1989. Although Islamabad claims it to be its indigenous production, there are enough reports which provide evidence of the Chinese assistance. Apart from China, certain European companies have also provided assistance to Pakistan.[27] There have been unconfirmed reports that the initial Hatf-1 and 2 project designs were both based upon a 1960s French sounding rocket design, known as Eridan.[28] Both Hatf missiles resemble the Chinese N series, so technical assistance from China cannot be denied.

Reports suggest that the Hatf-1 is a 70-km range unguided rocket, with a length of 6.0 m, a body diameter of 0.56 m and a launch weight of approximately 1,500 kg. The rocket is a single stage solid propellant system, with a payload of 500 kg that could be conventional High Explosive (HE), chemical or submunitions.[29] There have also been reports of the development of the improved version Hatf-1A, 100-km range missile. The first launching of the Hatf-1A took place on February 7, 2000. Subsequent launching of the improved versions of the Hatf-1 was reported to have taken place on May 28, 2002, March 26, 2003, March 31, 2005 and February 19, 2006.[30]

It is believed that the Hatf-1 entered service in 1992 and the Hatf-1A in 1995. The guided Hatf-1B, a further improved version with an upgraded kit fitted to the existing missile, was reportedly flight tested in February 2000 and is believed to have entered operational service in 2004.[31]

### *Hatf-2 (Abdali)*

The original version of the Hatf-1 SRBM was started in 1987, and was first deployed in 1989 as a two-stage version of the Hatf-1 missile. The Hatf-2 uses the Hatf-1 as a second stage, and has a range of 300 km with 500 kg payload.[32] There are unconfirmed reports that the Hatf-2 is an upgraded version of the Hatf-1B and was developed with Chinese aid and technical assistance. There was confusion regarding the name of the programme, and some reports suggest that it may have been referred to as Shadoz (King Hawk).[33] Although the production for the Hatf-2 started in 1987, the programme was reportedly delayed, in all probability due to the acquisition of the M-11 in the early 1990s. A new programme was later commenced in 1997 by the Pakistan National Development Complex (now NESCOM) with a reduced range of 180 km and was called the Hatf-2 or Abdali.[34] The Abdali missile was first flight tested in May 2002, and appears similar in size and shape to the Argentinean Alacran SRBM and the Chinese TY-3, TY-13 and TY-14 research rockets, which confirms the Chinese assistance in the build-up of the missile.[35] Due to their limited range, it is unlikely that the Abdali missiles can carry nuclear warheads. The missiles do not pose a significant strategic threat but their evolution does provide an insight into the development of Pakistan's missile programme and also missile technology transfer from China.[36]

### *Hatf-3 (Ghaznavi)*

There has been much speculation about the origin of the Hatf-3 Ghaznavi missiles. The original plan for the development of the Hatf-3 started in 1987. The original version was a two-stage missile using the design of the Hatf-2 which was planned to be modified to give it an extended range of 800 km. The programme for the development of the Hatf-3 was terminated after Pakistan received the M-11 from China. The technology for the M-11 was used to develop future missiles—the Shaheen-1 and Shaheen-2—in Pakistan. But, as a low priority, the programme for the Hatf-3 - Ghaznavi was initiated in 1997 in Pakistan by NESCOM. The first flight test was made in May 2002.[37] The other three test flights for the Ghaznavi-3 took place in October 2004, November 2004 and December 2006.[38] There are reports suggesting that these missiles are operational. Technical evaluation of the missile images suggests that the Hatf-3 is a version of the M-11 or may even be a repainted M-11. Reportedly, the production facilities of these missiles have been set up with Chinese assistance.[39]

The Ghaznavi missile is 8.5 m long, has a body diameter of 0.8 m, and a launch weight of 4,650 kg. The missile has a single stage solid propellant motor providing it a minimum range of approximately 50 km, a maximum range of 280 km, and it can carry single warhead of 700 kg.[40] The M-11 Chinese missile now has a range of 350 km and, thus, there are apprehensions that perhaps the Ghaznavi also has a range of up to 350 km.[41] The Ghaznavi can carry two kinds of warheads – a longer nuclear warhead and a shorter conventional warhead. Studies suggest that the warhead dimensions closely match the warheads of the Chinese M-11missiles.[42]

### *Hatf-4 (Shaheen-1)*

The Hatf-4 or Shaheen-1 is believed to be a scaled up version of the M-11 missiles supplied to Pakistan in 1993. Since 1996, there have been suggestions that a solid propellant missile was being developed by Pakistan (obviously with Chinese assistance). Ground tests of the motor for this missile were reported in 1997 and 1998, and it was first displayed in 1999.[43] Reportedly, the Shaheen-1 was developed by the Pakistan National Development Complex (PNDC) possibly with assistance from SUPARCO and the PAEC.[44]

The Shaheen-1 is a single stage, solid propellant missile with an inertial guided system and a maximum range of 750 km.[45] Two flight tests were made in October 2002, two in October 2003, one in December 2004, and one in November 2006. The Shaheen-1 was officially handed over to the Pakistan Army Strategic Missile Group in March 2003.[46] Reportedly, the warheads of the M-9 and Shaheen-1 are similar and a comparison of the warhead parts of the two missiles confirms a common origin.[47]

### *Hatf-5 (Ghauri)*

Pakistan's second most crucial partner in the missile development programme has been North Korea, while Pakistan, has been extending military assistance to North Korea in return. B. Raman has aptly said: "North Korea's assistance to Pakistan in the development of its missile capability has been as a *quid pro quo* for the latter's assistance to North Korea in the development of its military nuclear capability."[48]

It is believed that one of the nuclear devices tested in 1998 at Chagai was of North Korean origin.[49] Pakistan's link with North Korea was established in 1993 during the second tenure of Benazir Bhutto. The Pakistani Prime Minister visited Pyongyang after having talks with North Korean President Kim Il Sung. She was accompanied by Foreign Minister Sardar Asef Ahmad Ali and Minister of Defence Aftab Shaban Mirani, which clearly indicated the agenda for the visit. According to Pakistani officials, the Pakistani delegation went with plans for North Korea's No-dong missile.[50] While Pakistan's clandestine missile transactions with China have been closely monitored by the US, the same does not stand true for its clandestine missile technology transfer from North Korea which has enjoyed greater leverage. The US intelligence community has been warning the US Administration about the North Korea–Pakistan links related to missile transfer since 1997.[51]

The development programme for the Hatf-5 or Ghauri commenced in 1993 at the Khan Research Laboratories and was publicly announced in 1997. An engine test was carried out in January 1998, and the first flight test was made in April 1998, although several motor tests were carried out in North Korea in 1997 that might have been connected with the Ghauri.[52] Although North Korea has consistently denied providing assistance, the US intelligence sources did monitor the North Korean flights to Pakistan. The flights mostly involved IL-76 transports; the transports apparently carried technical experts, including telemetry crews.[53] The number of North Korean flights increased and this was followed by the visit of North Korean Chief of Staff and head of the strategic forces, which was indicative of the reported fact that the two nations have entered into an agreement which allowed North Korea access to Pakistan's range facilities in exchange for military technology.[54]

Reportedly, the 5-10 No-dong missiles assembly sets were sent to Pakistan between 1994 to 1997 for trials and to set up a manufacturing capability.[55] The Ghauri missile, which resembles the Russian 'Scud B,' is an outcome of coordinated inputs from both North Korea and China. There were reports regarding an arrangement among Pakistan, China and North Korea whereby China would provide the soft technology and engineering for the Ghauri, and North Korea would act as an agent for the transfer of Chinese technology and provide the hardware and components from its No-dong missile programme.[56] It has been reported that China provided Pakistan with a nuclear warhead design with a weight of 500 kg in the early 1980s, and this design was used to provide the warhead for the Ghauri missile. Also, there have been unconfirmed reports that the Chinese guidance systems have been used, which could have been passed through North Korea to Pakistan.[57]

The first flight test of the Ghauri, single stage, liquid propellant missile, with a range of 800-1,200 km, was in April 1998. An improved version, known as the Ghauri-2, was reportedly under development in 1998, with a enhanced range of 1,500 to 1,800 km, and was first flight tested in April 1999. The Ghauri-3 programme has been under development since 1994, with a range of 3,000 km and the first stage motor tests were reportedly done in July and September 1999.[58]

### *Hatf-6 (Shaheen-2)*

There were unconfirmed reports regarding the development of the Shaheen-2, also designated as Hatf-6, which is a two-stage solid propellant ballistic missile. Some reports also suggested that it was ready for testing in June 1999.[59] However, two Shaheen-2 missiles were displayed during the Pakistan Day Parade in Islamabad on March 23, 2000.[60] One of the missiles was carried on a 12-wheel transporter erector launcher, while the other Shaheen was carried on a missile transporter. These vehicles are apparently much larger in size than the 8 wheel launcher used by the Shaheen-1.[61]

On March 9, 2004, Pakistan test-fired the nuclear capable Shaheen-2 ballistic missile. Samar Mubarakmand, Chairman, National Engineering and Science Commission made a statement that the missile was a two-stage rocket weighing 25 tons with a diameter of 1.4 m, length of 17.5 m, and a range of 2,500 km.[62] The Shaheen-2 is believed to be based upon the earlier Chinese two-stage solid propellant missile M-18, which was demonstrated in 1988. The maximum range of the Shaheen-2 missile was 2,000 km, which has now been increased to 2,500 km, sufficient to target any important part of India[63] (See Fig 4.1). The Shaheen-1 and 2 have been developed by the PNDC with the assistance of SUPARCO and PAEC. There are unconfirmed reports that a three-stage Shaheen-3 missile, with a range of 4,000 km, might be under development in Pakistan for dual use, as a ballistic missile as well as a satellite launch vehicle.[64]

### *Hatf-9 (NASR)*

On April 19, 2011, Pakistan tested its short range surface-to-surface multi-tube ballistic missile Hatf 9 (NASR). The official press release for NASR said:

> [The NASR Weapon System] has been developed to add deterrence value to Pakistan's Strategic Weapons Development programme at shorter ranges. NASR, with a range of 60 km, carries nuclear warheads of appropriate yield with high accuracy, [and] shoot and scoot attributes. This quick response system addresses the need to deter evolving threats.[65]

Although a missile of 60 km range is more likely to be a free flying rocket, Pakistan has claimed the missile to be nuclear capable, which is possible. In all probability the missile is a four-tube adoptation of a Chinese-design Multiple Rocket Launcher (MRL), possibly the A-100 type, on an eight-wheeler truck, capable of carrying four ready to fire ballistic missiles.[66] The diameter of the Hatf -9 is just about 12 inches, which indicates that most probably, Pakistan has developed or acquired nuclear warheads small enough to fit into these missiles. According to a Pakistani academic: "The diameter size of NASR suggests, that the warhead would be less than 1 kg and would be of sub-kiloton range, suitable for battlefield use or could be a fission boosted sub-kiloton fission device".[67]

The NASR provides Pakistan with short range missile capability in addition to the long range ballistic missiles and cruise missiles. Also, according to the Pakistani military officials, the Hatf 9 belongs to the category of Tactical Nuclear Weapons (TNW) and is a low yield battlefield deterrent, capable of inflicting damage on armed brigades and divisions.[68] According to Pakistan, the Hatf-9 is their counter to India's Cold Start doctrine which envisions limited conventional response from the Indian side in response to a sub-conventional attack on India originating from the Pakistani territory.

**Fig 4.1: Pakistan's Ballistic Missiles**

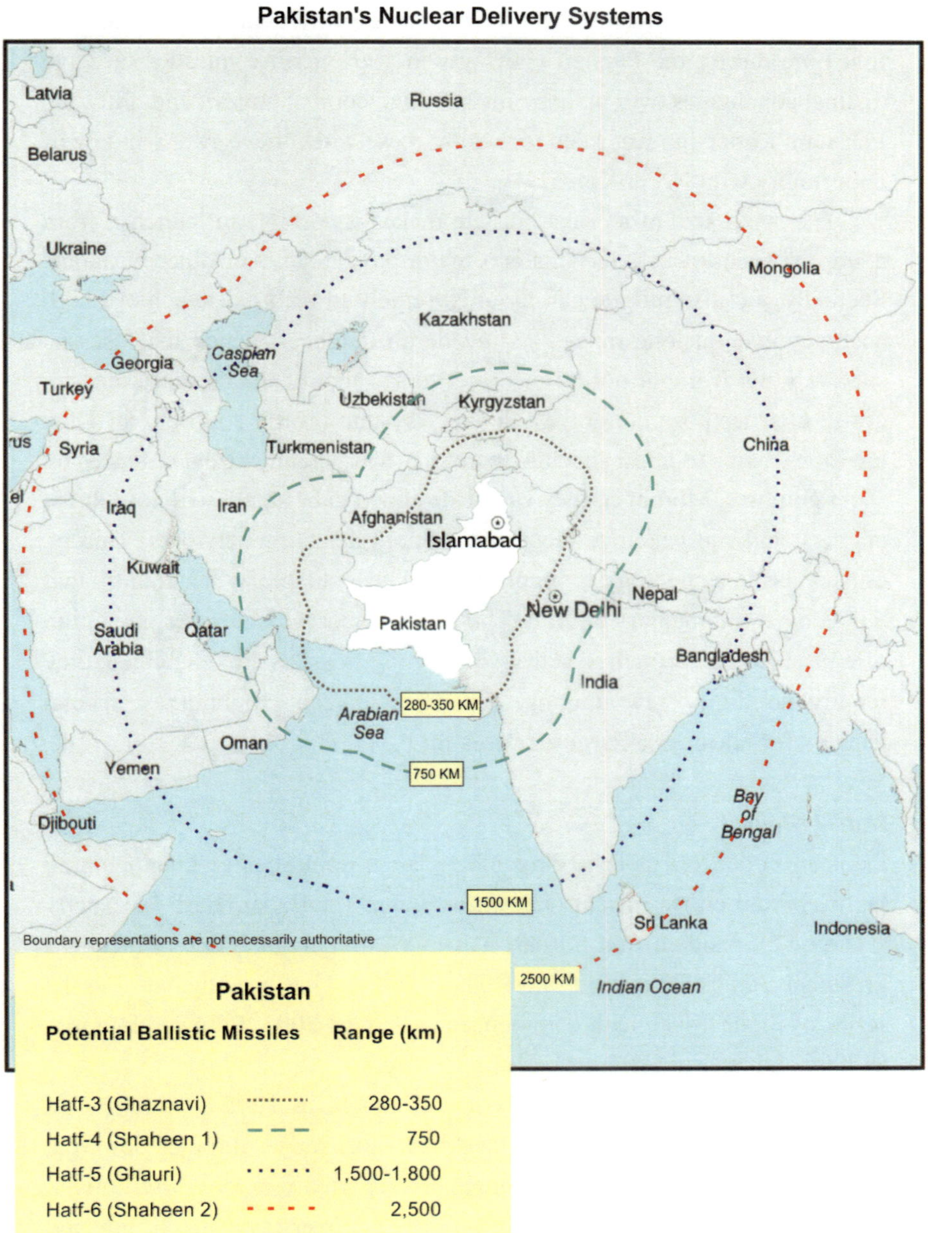


Boundary representation is not necessarily authoritative.

**Cruise Missiles**

The importance of cruise missiles is growing in the inventory of modern military forces. The US obviously pioneered the development of cruise missiles. The effectiveness of these missiles was demonstrated for the first time during the Persian Gulf War in 1991. Cruise missiles do have distinct advantages over ballistic missiles and combat aircraft and, thus, the Pakistani leadership was keen to acquire these when they saw a window of opportunity with North Korea.

The small size of a cruise missile makes it possible to launch it from a ship or even a truck and this is a major advantage over other weapons. Secondly, a cruise missile can fly at extremely low altitudes, achieve high accuracy even at long ranges and evade air defences, which increases its survivability. It might not be incorrect to say that in the current scenario, cruise missiles pose the gravest delivery system proliferation threat. They are inexpensive to build and can, therefore, overwhelm current defences by sheer numbers. Moreover, they can be designed to be small, with low-thrust engines, and can penetrate radar and infrared-detection networks. Thirdly, cruise missile technology is simple and can be available with a country that builds even rudimentary aircraft. The fourth factor which makes the option of cruise missiles attractive is that since cruise missiles are unmanned, they require no flight crew training, expensive upkeep programmes, special hangars for housing, or large air bases for basing.[69]

***Hatf-7 (Babur)***

Pakistan conducted a test-firing of its latest domestically manufactured (as it claims) cruise missile, known as Babur (or Babar, Hatf-7), exactly at the time President Asif Zardari was in Washington and due to meet US President Barack Obama, in May 2009.[70] This test has been the latest in the series of flight tests which commenced in August 2005, followed by more in 2006 and later in 2007.

Development of the Babur, also designated as Hatf-7, which is an air, ground, ship and submarine launched short range, turbojet powered, single warhead cruise missile, commenced reportedly in 1990 itself, given Islamabad's quest for enhancing its options of delivery systems. In July and

August 1998, two US RGM/UGM-109 Tomahawk cruise missiles were recovered almost intact in southern Pakistan, and it is believed that these may have been used for reverse engineering or even to contribute basic technology for the development of the Babur.[71] But this does not deny the Chinese technical input in the development of the Babur, as the missile looks similar to the Chinese Hong Niao-3, the US RGM-109 and also has similarities with the Russian SS-N-27 Club (3M14 version) cruise missile.[72] The total length of the missile, including the boost motor assembly, is believed to be 6.2 m, with the launch weight being around 1,200 kg; the payload is probably 450 kg, range 500 km and the warheads can be HE, either unitary or submunitions, or nuclear with a yield between 10 and 35 kT.[73] Although the first flight test was reported in August 2005, the production with NESCOM in all probability started in 2006.[74]

Various reports indicate that Pakistan is seriously undertaking efforts to upgrade the Babur and develop a new variant of this missile, the Babur-2, which would enhance its payload and range. It is developing an air-launched version of the Babur, which will reportedly be carried by F-16 and JF-17 aircraft and a sea launched version for deployment on the Agosta submarine.[75]

***Hatf-8 – Ra'ad***

In May 8, 2008, Pakistan tested a nuclear-capable, air-launched cruise missile, the Hatf-8 – Ra'ad, with a range of 350 km. The first test-launch for the Ra'ad was carried out in 2007. The Hatf-8 missile has been developed exclusively for launch from aerial platforms, enabling Pakistan to achieve a greater strategic capability on land and at sea.[76] Reports suggest that the Hatf-8, has special stealth capabilities and is a low altitude terrain following missile with high manoeuvrability, and can deliver all types of warheads, with high accuracy.[77] Unconfirmed reports suggest that the missle is possibly based on earlier versions of South African designs, such as the Raptor 2 (H-4) MUPSOW and Torgos.[78] In 2004, the Raptor-2 were exported to Pakistan.[79] The Ra'ad has a length of 4.85m, a square cross-section body, width and height of 0.53m, launch weight of 1,100 kg and the payload is approximately 400 kg. The payload could be HE unitary, 15 kT nuclear or submunitions.[80] Although the missile was initially launched from a PAF

Dassault Mirage III combat aircraft, it is planned to be integrated with, and launched from, other PAF platforms like the JF-17s, F-16s and (may be) the J-10s. (See Fig 4.2).

**Fig 4.2: Pakistan's Cruise Missiles**

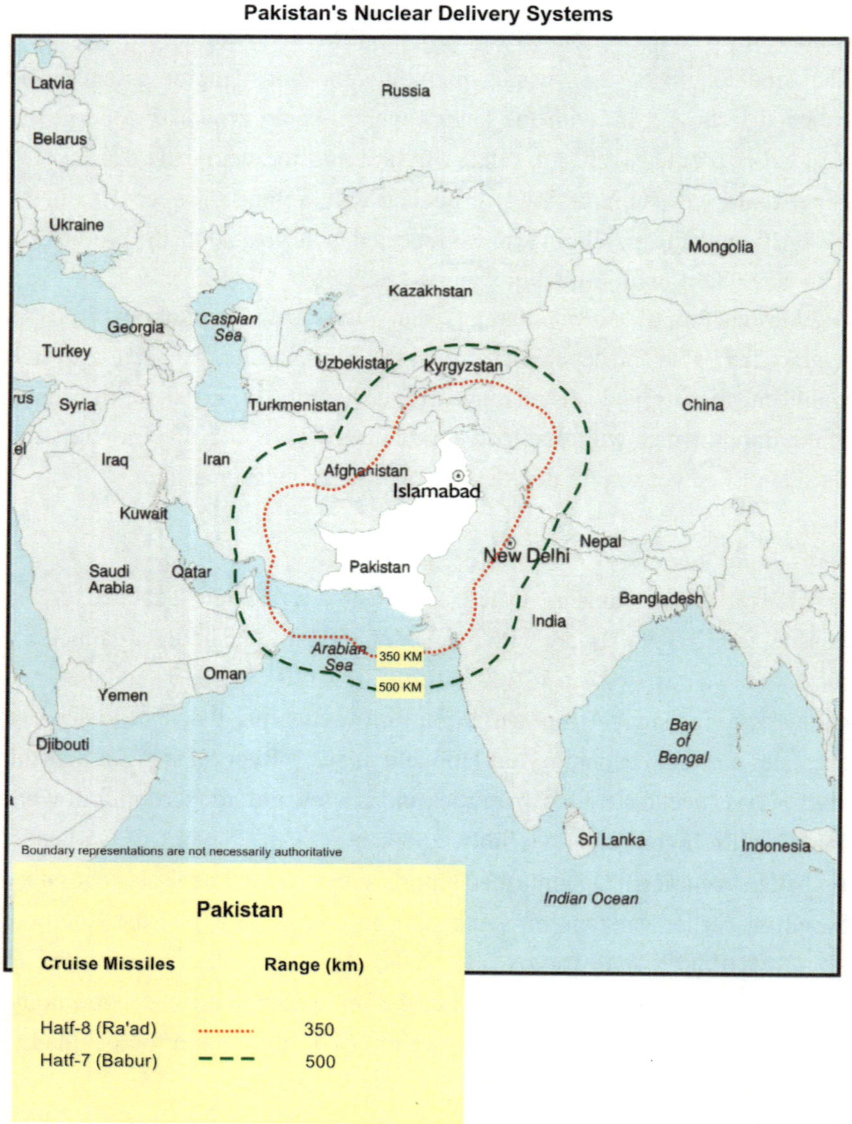


Boundary representation is not necessarily authoritative.

**Aircraft**

In the 1980s, Pakistan's nuclear development programme was in full swing, and aircraft were considered the chief means of delivering nuclear weapons. The main sources of its aircraft have been the United States, France and China. Pakistan's desire for strategic support from, and alliance with, the West to acquire high technology weapons to match the conventional military superiority of India has driven it towards the US. Nevertheless, Pakistan's alliance with Washington over the last 60 years has been shaped more by the American strategic requirements than the Pakistani needs and desires.

***Acquisition of the F-16s***

Pakistan openly announced its nuclear weapon capability in 1987 and, it can be presumed that in the early 1980s, it was clearly exploring opportunities to acquire the delivery systems when it entered into an alliance with the US in the war against Communism. Gen Zia-ul-Haq managed to negotiate a generous aid package of $3.2 billion (the military component of the aid package was worth $1.6 billion). Military assistance programmes from the US included the sale of 40 F-16 Falcon fighters/interceptors, one of the most advanced military aircraft in the world. In December 1981, Pakistan signed an agreement with the US for the purchase of 40 F-16s and according to the agreement, the deal would be split into two batches, one of 6 aircraft and the other of 34.[81] Pakistan did receive these 34 F-16s and 6 F-16A and F-16B from the US by the end of 1987.[82]

The second US aid package worth $4.02 billion for a period of six years, commenced in 1987, but was suspended in 1990 due to the arms embargo imposed under the Pressler Amendment. This was highly disappointing for Pakistan as it dried up the American supply of equipment to Pakistan. Although the Brown Amendment passed in 1995 permitted Pakistan to take possession of the military equipment frozen in the United States, it excluded the F-16s. What was extremely disturbing for Pakistan was that the US refused to export 28 F-16s which Pakistan had paid for (and which were also manufactured) against the 1988 order for the 110 F-16s. The US agreed to try and sell them to a third country and return the money to Pakistan and, thus, there were efforts to sell them to Indonesia in 1996, but the economic crisis stopped that plan. The Clinton Administration did face problems with respect to these 28 F-16s as President Clinton had pledged to the Pakistani Prime Minister, Ms Benazir Bhutto, that the money paid for the F-16s by Islamabad would be reimbursed in case the equipment was not delivered to Pakistan.[83] Reportedly, a number of countries, including Phillipines and New Zealand seriously tried to acquire these aircraft but the final deal did not work out.

However, post 9/11, the US aid and weapons sales were restarted following Pakistan's role as the chief ally in the global war against terrorism, which led to the removal of US sanctions on Pakistan. Pakistan's alliance with the US helped in the recovery of Pakistan's economy and opened up the supply of the long desired defence equipment from Washington, including the much awaited F-16s. Pakistan received the first two F-16s in 2005 and currently the total number of F-16s with the PAF stands at 59 (46 F-16A/F-16B + 1F-16C/D +12 on order).[84] However, certain unconfirmed sources report 30 plus aircraft on order. Pakistan has been extremely keen to acquire the F-16s partly because of the political/psychological reasons, and partly because it has been familiar with multi-role combat aircraft since the early 1980s. The issue has apparently been the cost and the level of upgradation of the aircraft which Pakistan would receive. Pakistan would aim to receive its original plan of 110 F-16s in its inventory, but much more capable ones.

The F-16 is a flexible design, capable of high performance in both the air superiority and ground attack roles. The flight controls are digital computer-controlled fly-by-wire, complemented by advanced navigation and avionics systems. The PAF deploys its F-16s with Squadrons 9, 11 and 4 at Sargodha Air Base, located 160 km northwest of Lahore.[85] The F-16 can carry up to 5,450 kg externally on one underfuselage centreline pylon and six underwing stations and has a refuelled range of more than 1,600 km. Given that the F-16 is undoubtedly the most capable Pakistani attack aircraft, it would likely be tasked with the delivery of nuclear air-to-ground munitions.

### *The JF- 17s and J-10s*

Sino-Pakistan defence collaboration flourished under the umbrella of the US sanctions (particularly in the 1990s) and, in the process, the two nations entered into deals for the co-development of a fourth generation fighter aircraft, the JF-17 (earlier called the FC-1). China delivered 12 Joint Fighter (JF-17) Thunder advanced jets to the Pakistan Air Force between 2007-09 for flight tests and evaluation. The JF-17 is designated to be a low cost, multi-role combat aircraft to meet the tactical and strategic requirements of the Pakistan Air Force, thereby reducing the country's reliance on imports. The JF-17 is co-developed by Pakistan and China and is being built by China's Chengdu Aircraft Industry Corporation (CAC) and Pakistan Aeronautical Complex (PAC), Kamra. There have been reports that the design was developed by the MiG complex in Russia and transferred to China after the Russian Air Force cancelled procurements. Pakistan has also increased its initial target of buying 150 JF-17s to acquiring up to 250 aircraft.[86] This represents a quantum jump in the Pakistani aircraft industry. Reportedly, the co-production of sub assemblies and structural parts has commenced and Pakistan is expected to join the exclusive club of the few nations manufacturing fighter aircrafts.

The FC-1 (JF-17) is fairly flexible in terms of avionics and weapon configurations. The JF-17 in service in the PAF is presumably fitted with the Italian Grifo S-7 fire-control radar. The radar has 25 working modes and a non-breakdown time of 200 hours. Further capabilities include look-down, shoot-down and ground strike abilities.[87] The aircraft is fitted with PL-12 or SD-10 radar homing Medium-Range Air-to-Air Missiles (MRAAM), anti-ship missiles, anti-radiation missiles, unguided bombs rocket launchers and a laser guided pod.[88] The aircraft is equipped with AIM-9L/M, PL-5E, PL-9C short range air-to-air missiles. It can carry up to 3,700 kg payload and the range and service ceiling of the aircraft are 2,037 km and 15,240m respectively. The FC-1 being produced for the PAF will be fitted with Inflight Refuelling (IFR) probes.[89]

Pakistan is also positioning itself to buy up to two squadrons of the Chinese J-10s which, along with the JF-17, would form the backbone of the PAF, according to the Pakistan Air Force Chief. The J-10 (Jianji-10, or Jian-10) is an all-weather multi-role fighter aircraft designed for both air-to-air and air-to-ground missions.[90] The J-10 was designed by the Chengdu Aircraft Design Institute (611 Institute) and built by the Chengdu Aircraft Corporation (CAC) and has been operational with the PLA Air Force (PLAAF) since 2003.[91] The J-10 programme began in the mid-1980s as an air-superiority fighter to rival the then emerging fourth generation fighters such as the F-16 and MiG-29. However, the end of the Cold War and the changing requirements shifted the development towards a multi-role fighter with both air-to-air and ground attack capabilities.[92] The programme was assisted by Israel, which apparently provided the "technologies for its IAI Lavi lightweight fighter including the aerodynamic design and the software for the fly-by-wire flight control system."[93] In the early 1990s, the development programme faced several difficulties due to the arms embargo on China, imposed by the United States and European Union. In the mid-1990s, Russia got involved in the development programme and China received the Russian Al-31F turbofan jet engine to power the aircraft. Thus, the Chinese benefitted from both the Israeli and the Russian technology in the development of the J-10.[94]

The J-10 marks one of the most significant achievements of the Chinese aviation industry in the 1990s. For air-superiority missions, the aircraft can carry a mix of MRAAM and Short-Range Air-to-Air Missiles (SRAAM).The J-10 was designed with the surface attack capability right from inception.[95] The J-10 programme was kept under tight security and high secrecy. The Chinese state media only announced the J-10 in November 2006, nearly two years after it entered service. Despite the huge publicity the J-10 has enjoyed in the Chinese media, no official data has been provided regarding the actual capabilities and performance of the aircraft. Both the JF- 17 and J-10 are believed to carry nuclear warheads.

### *A-5s and French Mirage Vs*

Pakistan acquired 60 A-5s from China, and as of mid-1999, only 49 remained in service. Reportedly, some of the A-5s are capable of carrying nuclear bombs. But given their payload capability, the bomb would have to be quite small. The aircraft offers enhanced combat performance, particularly at low and super-low altitude.[96]

Pakistan could also potentially use the French Mirage Vs for the nuclear-strike mission. Technically speaking, the Mirage Vs could be deployed as part of the 8th (Haider) Squadron of the 32nd Fighter Wing at the Masroor Air Base, located about 8 km west of Karachi.[97] They could also be deployed as part of the 25th (Eagles) Squadron of the 33rd Fighter Wing of Kamra Air Base located 65 km west of Islamabad.[98] The nuclear capable cruise missile Raad–Hatf-8 could be potentially deployed to the Mirage V squadrons in the future.

**Fig 4.3 : Pakistan Air Force Bases**

Boundary representation is not necessarily authoritative.

*Source*: Modified version of map taken from http://www.globalsecurity.org

**Table 4.1: Pakistan's Nuclear Delivery Systems**

| Type | Range (km) | Payload (kg) |
|---|---|---|
| **Ballistic Missiles** | | |
| M-9 | 600 km | |
| M-11 | 300 km | 800 kg |
| Hatf-3 (Ghaznavi) | 280-350 km | 700 kg |
| Hatf-4 (Shaheen 1) | 750 km | |
| Hatf-5 (Ghauri) | 1,500-1,800 km | |
| Hatf-6 (Shaheen 2) | 2,500 km | |
| Hatf-9 (NASR) | 60 km | |
| **Cruise Missiles** | | |
| Hatf-7 (Babur) | 500 km | 450 kg |
| Hatf-8 (Ra'ad) | 350 km | |
| **Aircraft** | | |
| F-16A/B | 1,600 km | 1 bomb |
| F-16C/D | | |
| Mirage V | 2,100 km | 1 bomb |
| A-5 | | |
| JF-17 | | |
| J-10 | | |

*Source: Jane's Strategic Weapon Systems,* Issue 48, 2008; *Jane's All the World's Aircraft*, various issues.

### Implications for India

Pakistan's missile delivery systems have been developed significantly, and it is in the process of rapid modernisation of its combat fleet, with a clear focus on building up the numbers as well as the quality of the aircraft. Thus, it might not be incorrect to state that Pakistan is capable of a nuclear exchange if the situation arises.

Pakistan's build-up of the delivery systems and specifically, the missile build-up, like the nuclear build-up, is India-centric. Pakistan has consistently claimed that it has been forced into the missile build-up given India's growing missile muscle, but the reality is that India's strategic requirements are far more complex and demanding as compared to Pakistan's. India needs to take the consistent China threat into consideration in its strategic and defence planning—a threat which does not exist for Pakistan. On the other hand, for Pakistan, China is the most trusted ally and has offered it military and diplomatic assistance against India in the previous wars. In fact, in any future conflict between India and Pakistan, there is no reason for India to presume that China would stay neutral. India strongly feels the need to prepare itself for a possible two-front war and, therefore, Indian defence requirements are far more complex.

Pakistan's belief in deterrence and increasing the options of delivery systems provides it with enhanced deterrence. There is a clear linkage between nuclear deterrence and ballistic missile capability. For Pakistan, the ballistic missile capability enhances its deterrence, and also its choices for preemptive strikes, given its "first use" nuclear doctrine.

The Chairman, Joint Chiefs of Staff Committee (CJCSC), Gen Tariq Majid, while addressing the convocation ceremony at the National Defence University, Islamabad said, "Our nuclear weapons are the cornerstone of Pakistan's deterrence doctrine and we are determined to retain it at all costs. No amount of coercion, direct or indirect, can force us to compromise on this core interest."[99]

Lt Gen F. S. Lodhi has spoken on the same lines:

> In the modern defence concept, the missile system is the most essential element. In fact, it is now the core of any viable defence structure and the cutting edge of an adequate defence capability of any nation. It cannot, therefore, be ignored by the defence planners. In Pakistan's security environment, an adequate missile defence will prove an effective and reliable deterrent. The essence of deterrence worldwide, is a country's power to retaliate in kind. It was, after all, the power of deterrence that prevented a third World War between the Western allies and the Soviet empire for over 50 years.[100]

Pakistan has been focussed on acquiring missiles with increasing range to diversify its targets (Table 4.1). Pakistan's initially developed the SRBMs Hatf-1, Hatf-2, Hatf-3 which can be used against military targets such as bases or airfields. These carry relatively smaller warheads and, thus, it would not be practical for Pakistan to deploy these against civilian population centres. The Hatf-5, which is an Intermediate Range Ballistic Missile (IRBM), has a range of up to 1,800 km (the improved version would have a range of approximately 3,000 km) and can carry a 750 kg payload. Its main drawback is its limited accuracy and, thus, its effective use would be striking civilian targets. The Hatf-5, according to the Pakistani leadership, would give them an advantage over India's strategic assets as some of the important Indian cities, power plants and water purifiers would fall in this range (Fig 4.1). According to Maj Gen F.S. Lodhi (Retd):[101]

> When Hatf V (Ghauri) missile is deployed in Pakistan, it will cover most of India except its eastern and southern portion and the coast, along the Bay of Bengal. It will deprive India of the advantage of strategic depth that it enjoys at present in relation to Pakistan. It will, therefore, give Pakistan a degree of defence parity that has been reducing in the last few years. Pakistan will now be in a position to hit back effectively if subjected to aggression by India and inflict unacceptable damage to India's important and vulnerable areas and particularly those areas which were hitherto considered safe, owing to the distance from the Pakistan border.

Testing of the Pakistan's latest missile Hatf-9 (NASR) in April 2011 was advertised as an answer to India's limited war options, especially against an incursion by land forces. But, limited war can be effectively dealt with by the conventional weapons also. The use of nuclear weapons in any form (even if they are tactical nuclear weapons) by Pakistan, would face an Indian response based on the Indian nuclear doctrine which states: "Any nuclear attack on India and its forces shall result in punitive retaliation with nuclear weapons to inflict damage unacceptable to the aggressor."

Essentially, the Hatf-9 does not provide an answer to Pakistan and it has its own downsides. Possession of tactical nuclear weapons with an objective of a quick response would certainly increase the risk factor involving the use of nuclear weapons by miscalculation or even nuclear weapons falling into unauthorised hands.

The growing missile muscle and also the enhanced number of nuclear capable combat aircraft would increase Pakistan's offensive capabilities against India. In the previous decades, Pakistan has been deterred by India's superior conventional military build-up and has consistently used it as an excuse for demanding high technology weapons from the West. With the current acquisitions plans, the Pakistan Air Force would grow tremendously in the next 5-15 years and the multi-role combat aircraft which comprised 11 percent of the PAF inventory, would go up to as much as 42 percent, with more F-16s, and the J-F17s and the J-10s coming in. The Chief of the PAF, in March 2009, said, "We have made the whole of PAF a nuclear force," with the aim of giving it the status of a real deterrent force.[102]

Fear of risk and punishment is central to the concept of deterrence and with this growing nuclear arsenal, Pakistan will eventually lose the fear. In fact, it might not be incorrect to say that post Parakaram, Pakistan, by and large, believes it would not face Indian retaliation, and can continue abetting terrorism on Indian soil.

## Notes

1. Ann Tibbitts Schulz, *Buying Security:Iran Under the Monarchy* (San Francisco: Westview Press, 1989), p.54, as cited in "Pakistan Profile: Missile Chronology", *NTI* at http://www.nti.org/e_research/profiles/Pakistan/Missile/3068_3069.html
2. Brahma Chellaney, "Missile Muscle in the New Definition of World Power", *The Asian Age*, April 16, 1999.
3. Dr Subhash Kapila "Pakistan's Ballistic Missile Arsenal: Development and Acquisition", South Asia Analysis Group Papers, at http://www.southasiananalysis.org/%5Cpapers2%5Cpaper148.html
4. Ibid.
5. There is no doubt that this would have created a major rift between the US and Pakistan, which may have been Mao's real motive.
6. Roger Frost, "Pakistan's Procurement Chief on New Programs", *International Defence Review,* June 1989, p. 765.
7. Bill Gertz, "China Plays a Major Role in Expanding the Nuclear Club", *Washington Times,* May 29, 1998.
8. "Nuclear Weapons Database: Pakistan's Possible Nuclear Delivery Systems", at http://www.cdi.org/nuclear/database/panukes.html

9. "Weapons of Mass Destruction: DF-11 [CSS-7], DF-11A", at http://www.globalsecurity.org/wmd/world/china/df-11.htm
10. Bill Gertz, *Washington Times*, October 4, 1994, p. A8; Barbara Starr, *Jane's Defence Weekly*, October 15, 1994, p. 6, as cited in "China's Missile Exports and Assistance to Pakistan –Statements and Developments" at http://cns.miis.edu/archive/country_india/china/mpakchr.htm
11. "Missile Overview", *NTI* at http://www.nti.org/e_research/profiles/Pakistan/Missile/index_3066.html
12. Bill Gertz, "China Can't Say No to Arms Buyers", *Washington Times,* May 28, 1991
13. Bill Gertz, "Missile Deception," in *Betrayal: How the Clinton Administration Undermined American Security* (Washington, DC: Regnery Publishing, Inc., 1999), p. 159, as cited in "Pakistan Profile: Missile Overview", *NTI* at http://www.nti.org/e_research/profiles/Pakistan/Missile/index.html
14. "Although the DF-11 has a range of 300 km, the Chinese continued work on a version with a longer range. China's 50th anniversary military parade on October 1, 1999, marked the first public Chinese display of a new version of the M-11 short-range missile, the CSS-7 Mod 2, more commonly known as the M-11 follow-on. The new Mod 2 missile is about two metres longer than the Mod 1, and believed to have a longer range, a larger warhead and greater accuracy than the earlier M-11. The accuracy of these missiles will improve in the future if China is able to apply Global Positioning System (GPS) guidance technology to provide highly accurate location information for missile launchers or pre-surveyed launch sites," cited in n. 9
15. Bill Gertz, "Pakistan Deploys Chinese Missiles", *Washington Times,* June 12, 1996.
16. Ibid.
17. "China's Missile Exports and Assistance to Pakistan", *NTI*, at http://www.nti.org/db/china/mpakpos.htm
18. Gertz, n.15.
19. Sanctions had a significant impact on the sales of high technology goods to China. Those goods were a major portion of the $12 billion US trade with China in the 1992.
20. Ibid.
21. Leon Hadar, "Report of Missile Export Deals Blow to China's Trade Status Extention," Bu*siness Times* (Singapore), May 8, 1993.
22. "Pakistan Denies Receiving Chinese M-11 Missiles," Agence France Presse, May 21, 1993.
23. Jeffery Parker, *Reuters*, December 1993.
24. Statement by Gordon Oehler, Former Special Assistant to the Director, CIA and

Director DCI's Nonproliferation Center; Hearing of the Senate Foreign Relations Committee on Proliferation of Chinese Missiles; Gary Milhollin, Director; Wisconsin Project on Nuclear Arms Control, June 11, 1998

25. "DongFeng 11 (CSS-7) Short-Range Ballistic Missile", at http://www.sinodefence.com/strategic/missile/df11.as p
26. Office of the Secretary of Defence, "Proliferation: Threat and Response," November 1997 (online version); US Department of State, Daily Press Briefing, August 26, 1996; R. Jeffrey Smith, "China Linked to Pakistani Missile Plant," *The Washington Post*, August 25, 1996, p. A1, A25; Tim Weiner, "US Suspects China is Giving Pakistan Help with Missiles," *New York Times*, August 26, 1996, p. A4, as cited in "China's Missile Exports and Assistance to Pakistan", at http://cns.miis.edu/archive/country_india/china/mpakpos.htm
27. *Jane's Strategic Weapons System*, Issue 48, 2008, p. 108.
28. Ibid.
29. Ibid.
30. S. Chandrashekar, Arvind Kumar and Rajaram Nagappa. "An Assessment of Pakistan's Ballistic Missile Programme: Technical and Strategic Capability", *NIAS* Study, 2006, p. 7.
31. n. 27, p. 109.
32. Ibid.
33. "Nuclear Weapons Database: Pakistan's Possible Nuclear Delivery Systems," at http:www.cdi/nuclear/database/panukes.html
34. n. 27, p. 109.
35. Ibid.
36. Chandrashekar et. al., n. 30, p. 9.
37. n. 27, p. 111.
38. Chandrashekar et. al., n. 30 p. 10.
39. Ibid.
40. Ibid.
41. Ibid.
42. Ibid, p. 13.
43. n. 27, p. 112.
44. Ibid.
45. Ibid.
46. "Pakistan Conducts Third Missile Test in 3 Weeks", *Rediff NEWS*, at http://in.rediff.com/news/2006/dec/09hatf.htm
47. Chandrashekar, et. al. n. 30, p. 24.
48. "Pakistan and the North Korea Connection", *Asia Times Online*, at http://www.atimes.com/atimes/South_Asia/DJ22Df01.html

49. Ibid.
50. "Bhutto Ends Visit to North Korea", Agence France Presse, December 30, 1993, International News; in Lexis-Nexis Academic Universe, December 30, 1993, web.lexis-nexis.com, at "NTI: Country Overview: Pakistan Missile Chronology" at, http://www.nti.org/e_research/profiles/Pakistan/Missile/chronology_1992_1993.html
51. In 1997, *Jane's Defence Weekly* reported that Gordon Oehler, former Director of the CIA's Nonproliferation Centre said that Pakistan has recently announced the development of a 1,500-km range missile called Ghauri. Even though Pakistan claims the Ghauri to be an indigenously produced missile, US analysts believed that China and North Korea provided the necessary technical advice
52. n. 27, p. 113.
53. "Pakistan's Missile 'Was a No-dong'", *Jane's Missiles &Rockets* (Coulsdon, Surrey), Vol. 2, No. 5, May 1998, pp.1-2.
54. Ibid.
55. Ibid.
56. Joseph Bermudez, "A Salient Partner", *Jane's Defence Weekly* (Coulsdon, Surrey), May 20, 1998.
57. n. 27, p. 113.
58. Ibid., pp. 113-114.
59. Ibid., p. 115.
60. "Shaheen-II / Eagle-I / Hatf-6 / Ghaznavi, Weapons of Mass Destruction (WMD)," at http://www.globalsecurity.org/wmd/world/pakistan/shaheen-2.htm
61. Ibid.
62 Ibid.
63. n. 27, p. 115.
64. Ibid.
65. Rodney W. Jones, "Pakistan's Answer to Cold Start?", *The Friday Times,* at http://www.thefridaytimes.com/13052011/page7.shtml
66. Ibid.
67. "Pakistan's Nuclear Artillery?", *Arms Control Wonk,* December 12, 2011.
68. "Pakistan Successfully Test-Fires Nuclear Capable Hatf-9", *The Express Tribune,* April 20, 2011.
69. "Cruise Missiles", Federation of American Scientists, at http://www.fas.org/nuke/intro/cm/index.html
70. "Pakistan Test-fires Babur Cruise Missile" at http://forum.pakistanidefence.com/lofiversion/index.php/t82549.html
71. n. 27, p. 117.
72. Ibid.

73. Ibid.
74. Ibid., p. 118.
75. "Pakistan Successfully Test-Fires Hatf-VII Missile", *Pak Tribune*, July 26, 2007. Hali, "Second Strike Capability", *The Nation* (Islamabad), August 16, 2006, as cited in *SIPRI Yearbook 2009* (Oxford: Oxford University Press, 2009), p. 375.
76. Iftikhar A. Khan"Cruise Missile Fired from Aerial Platform", *Dawn*, May 9, 2008, at http://www.dawn.com/2008/05/09/top4.htm
77. Ibid.
78. *Jane's Strategic Weapons Systems*, Issue Fifty-five, July 2011, p. 129.
79. Ibid., also see p. 184.
80. Ibid., p. 129.
81. "Pakistan Fiza'ya Pakistan Air Force-PAF", F-16.net The Ultimate F-16 Reference at, http://www.f-16_users_articles14html
82. Ibid.
83. Ibid.
84. *The Military Balance 2010,* IISS (London: Routledge, 2010), p. 369.
85. For details, see "PAF Mushaf / PAF Sargodha 32° 03' 09"N 72° 40' 07"E" at, http://www.globalsecurity.org/military/world/pakistan/sargodha-ab.htm
86. Interview, Air Chief Marshal, Tanvir Mahmood Ahmed, Pakistan's Chief of Air Staff, *Jane's Defence Weekly*, April 4, 2007, p. 34.
87. "FC-1 Xiaolong/JF-17 Thunder, China/Pakistan", at http://www.airforce-technology.com/projects/fc1xiaolongjf17thund/
88. Ibid.
89. Ibid.
90. "Jian-10 Multirole Fighter Aircraft", at http://www.sinodefence.com/airforce/fighter/j10.asp
91. Ibid.
92. Ibid.
93. Ibid.
94. Ibid.
95. Ibid.
96. "Fantan a-5, Q-5 (NANCHANG)" at http://www.globalsecurity.org/military/world/pakistan/a-5.htm
97. Robert S. Norris and Hans Kristensen, "Nuclear Notebook: Pakistani Nuclear Forces, 2009", *Bulletin of the Atomic Scientists*, September/October 2009, pp. 82-88.
98. Ibid.
99. "Nuclear Weapons are the Cornerstone of Pakistan's Deterrence Doctrine", Islamabad, June 18, 2009, *Pakistan Defence,* at http://www.defence.pk/forums/wmd-missiles/28455

100. Lt Gen Sardar F. S. Lodi(Retd), "Pakistan's Missile Technology", *Defence Journal,* May 1998, at http://www.defencejournal.com/may98/pakmissiletech.htm
101. Ibid.
102. "Whole of PAF Made Nuclear: Chief of Air Staff Air Chief Marshal Tanvir Mehmood Ahmed", *Asian Defence,* March 17, 2009, at http://theasiandefence.blogspot.com/2009/03/whole-of-paf-made-nuclear-of-air.html

# 5. Nuclear Doctrine and Strategy

After the overt nuclearisation in 1998, Pakistan pronounced some notions regarding its nuclear thinking which form the basis of its doctrine and strategies. Doctrine does acquire significance in the context of Pakistan's nuclear programme given its clandestine nature and lack of empirical evidence to support critical prepositions. Although Pakistan has consistently claimed that it developed its nuclear programme in response to security threats from India, more specifically, Pakistan traces the genesis of its programme to India's peaceful experiment in Pokhran, in 1974. But the fact is that Bhutto was seriously thinking of nuclear weapons much before India's peaceful experiment which became an immediate excuse for Islamabad. Pakistan's nuclear weapons are certainly not in response to India's nuclear weapons programme, but because Pakistan's ruling elites believe that nuclear weapons are the only means to match India's conventional military superiority. Zulfiqar Ali Bhutto had openly announced Pakistan's ambition for the nuclear bomb in January 1972 (well before May 1974). Pakistan has subsequently endeavoured to use the nuclear weapons to carry on, and intensify, its proxy war in Kashmir, claiming the Valley to be a "nuclear flashpoint". Pakistan has managed to pursue its grand strategy of "bleeding India through a thousand cuts" under the nuclear umbrella.

**Pakistan's Grand Strategy**

The nuclear strategy of Pakistan cannot be viewed in isolation and has to be analysed in the context of its "grand strategy". The military's grand strategy and, hence, Pakistan's strategy has been formulated and evolved over the decades on account of the perceived Indian threat, which has allowed the army to emerge as a "nation builder", legitimising the military's multi-dimensional role in the state. The military's grand strategy has rested on maintaining the centrality of the covert war (guerrilla war, war through terrorism) strategy. Pakistan's reliance on covert war through terrorism is unlikely to change in the coming years, although tactics and intensity may undergo changes. Similarly, its acquisition of nuclear weapons has been rationalised as a deterrent to Indian conventional military superiority and to provide an umbrella under which to pursue a proxy war through terrorism.

***Hostility Towards India and Search for Parity***

Pakistan's perceived security threat from India which emerges from its geographical and structural weaknesses prompted the nation to undertake its first aggression against India in 1947, and, for the second time, in 1965. Both times, Pakistan failed in its objectives which deepened its urge for a stronger military capability. India's conventional military superiority and Pakistan's desire to neutralise it has largely shaped the latter's defence build-up. The strategic aims, as brought out in the Pakistani writings, are: "to strengthen national power; to prevent open aggression by India; to induce India to modify its goals, strategies, tactics and operations; to attain a position of security or, if possible, dominance, which would enhance the role of other (non-military) means of conflict; to promote and capitalise on advances in technology in order to reach parity or superiority in military power."[1] Pakistan has relied on high technology weapons to seek competitive military advantage. The perceived military threat from India, which Pakistan considers as an "intelligent and implacable enemy"[2], has shaped Pakistan's decision to acquire and maintain technological superiority. "Pakistan must have a counter system for every Indian system, either to defend or to deter through the threat of riposte. If access to foreign aid is not assured, Pakistan needs to develop, and keep, a technology base sufficient to allow it to generate counter-systems to any new weapons the enemy might acquire through import or indigenous development"[3].

The desire to acquire high technology weapons has been very strong in the Pakistan military and the alliance with the United States has provided Pakistan with opportunities to acquire such weapons and systems. Pakistan believes that acquisition of high technology weapons would boost the morale and capability of the forces and, hence, improvement of the technological base and acquisition of advanced weaponry is vital for victory in war.[4]

***To Take Control of Kashmir and Revenge for 1971***

The need to control J&K and incorporate it into Pakistan was felt very strongly in Pakistan since the British declaration of India's partition. Kashmir was of utmost importance to both India and Pakistan not only because it was the largest princely state but, more because of its geographical location. Although the initial objectives behind Pakistan's aggression in the Valley in 1947 were stated as, firstly, security and, secondly, economic and psychological, the factor of 'religion' was added later on for justification for the aggressive actions.

Pakistan's fear could be very well sensed in Maj Gen Akbar Khan's writings. According to him, with J&K as an integral part of India, Pakistan's "security would be jeopardised if the Indian troops came to be stationed along Kashmir's western border".[5] The second factor was economic and psychological. Kashmir going to India would seriously impact Pakistan's economy which was based mainly on canal irrigated agriculture. West Punjab particularly depended for its agriculture upon the rivers flowing from J&K. Four out of the five rivers in West Pakistan, the Indus, Jhelum, Chenab and Ravi, originate from across the mountain reaches of Jammu and Kashmir, and the fifth river, the Sutlej (and Beas) combined flow through Himachal Pradesh, bordering J&K on its southern flank.[6]

Pakistan's concerns for security and about the waters of rivers originating from Kashmir were resolved as the result of the ceasefire agreement in 1949 and the Indus Water Treaty in 1960, respectively. But Pakistan's desire and struggle for Kashmir went on and, eventually, Kashmir became the excuse more than the cause of the conflict.

Pakistan's dramatic defeat in the 1971 War and its disintegration was indeed difficult for Pakistan to accept. Pakistan has never failed in accusing India for being responsible for the creation of Bangladesh. The loss of East Pakistan was Bhutto's political gain and it altered not only the political and ethnic scenario in Pakistan but also the ideological dynamics in Pakistan.

Pakistan's experience in the 1971 War led to two major developments in the grand strategy. First, the division of Pakistan created a fear of further segmentation, and religious ideology came out as the best choice for

Pakistan to unite the people in the name of Islam. The military and political leadership in Pakistan was totally convinced to follow the religious ideology to counter any further division of Pakistan and also to motivate the nation for an aggressive posture against India.

Second, India's victory in 1971 clearly indicated that it was difficult for Pakistan to win a conventional armed conflict with India, and, thus, going nuclear was seen by the leadership in Pakistan as the only solution. Bhutto, in his book, *Myth of Independence*, stated, "Pakistan's security and territorial integrity are more important than economic development" and "the defence requirements of her sovereignty have to be met first".[7] In the efforts to gain international support for the nuclear programme and provide it a legitimate base, Bhutto in 1974, hosted an Islamic Summit in Lahore urging the Muslim nations like Libya and Saudi Arabia to finance Pakistan's nuclear programme and, hence, the *Islamic bomb*. Bhutto propagated the religious ideology to seek assistance for Pakistan's nuclear programme on the global platform. Nuclear weapons were expected to "neutralise" India's conventional military superiority and, hence, increase Pakistan's military and psychological capacity to continue its fight for Kashmir.

***Covert War to Compensate for Imbalance in Conventional Superiority***

Pakistan adopted the route of covert war in India to make up for the deficit in defence capability. The strategy of covert war has been a component of the grand strategy from the beginning, and in 1947, when Maj Gen Akbar Khan wrote the plan for the first aggression under the title of "Armed Revolt Inside Kashmir". The *aggression* which is often termed as a 'revolt' did not break out of major dissatisfaction with the ruling class or the will of the Muslim population to seek Kashmir's accession to Pakistan. The aggression was a well thought out planned conspiracy by the Pakistani political leaders, Pakistan Army and ex-Servicemen of the erstwhile Indian Army and the Indian National Army, which, with the help of the local tribes, led to the outbreak of the 1947 War between the two newly independent nations.

In the 1965 War, the strategy of covert war was developed around Operation Gibraltar as a covert component. The operation was a failure as Pakistan in its planning overestimated the support of the locals in Kashmir.

Pakistan's covert activities continued in the 1970s and then increased in the 1980s in Jammu and Kashmir. Added to terrorism in the Valley, was Pakistan's support to terrorism in Punjab during 1983-93. Pakistan's covert war strategy was not limited to Islamic ideology and its religion. It exploited fundamentalist extremists among the Sikhs—Khalistanis—as the fighters for its covert war in Punjab starting in the mid-1980s. By 1990, there was a major expansion in the violence both in terms of numbers and intensity, and the killings in Punjab reached savage levels. The total number of fatalities in the Khalistan War went up from 29 in 1981 to 5,265 in 1991.[8] It cannot be denied that the Pakistani covert war strategy which aimed at reaching the northern part of India was keeping the Indian security authorities extremely busy. It was able to tie down a large number of military and police personnel in Punjab.

Pakistan's clear objectives in supporting terrorism in Punjab were:

- Pakistan's claim for Kashmir tempted the Inter-Services Intelligence (ISI) to encourage militancy in Punjab (especially west of the river Beas) as, according to Pakistan, this would make the Indian defence in Kashmir difficult, both by threatening the logistics lines of communication to J&K and also creating insecurity in the core area from where the Indian Army could launch a counter-attack to relieve any Pakistani military pressure in the Chhamb sector (as, indeed, was the case in the 1965 and 1971 Wars).
- To create political unrest in India which, in turn, would create instability and, thus, place Pakistan in a better position to conduct a war in Kashmir.
- Punjab has been an economically sound state and, thus, disturbances in Punjab would impact the overall Indian national economy.
- Given the size and diversity of India, the probability of the separatist movement in the heart of India could have spiralling effect and create the opportunity for the birth of the demand for a separate state from other regions.

In the late 1980s and 1990s, Pakistan became much more active in sponsoring terrorism in J&K. The ISI encouraged young Kashmiris to come to Pakistan for training. The ISI initially trained the secular groups in Kashmir and eventually shifted to training of the groups linked to Pakistan's own Islamic parties.[9] Pakistan's strategy of covert war became more concrete in the late 1980s and 1990s. And in 1999, India once again faced a Pakistani aggression where once again Pakistan initiated the conflict with its army in Kargil, while trained Mujahideen continued terrorism in the Valley and Jammu Division of J&K.

Pakistan has extensively used covert war to support the overt war whenever it deemed it necessary. While the strategy of covert war has been used by many countries (the US used it successfully against the Soviets during the Afghan War), Pakistan is the only nation to use covert war in combination with a full-fledged conventional confrontation.

### *Nuclear Weapons to Neutralise India's Conventional Superiority and Ensure Own Security*

The central assumption on which Pakistan has progressed and built up its nuclear arsenal is that a credible nuclear deterrent would compensate for the inferiority of its defence forces. According to Hasan Askari Rizvi, Pakistan lacks well trained, adequately equipped and numerically sufficient armed forces vis-à-vis India.[10] The basic rationale for Pakistan acquiring nuclear weapons has been its expectations to neutralise India's perceived conventional military superiority and the way it was employed by it in the 1971 War. Former Foreign Minister Agha Shahi referred to it as the "Sword of Damocles" hanging over Pakistan's head, when stating the objectives of Pakistan's nuclear weapons. They were, he said, "...to equalise, to compensate our military imbalance that hangs like a Sword of Damocles over the head of the nation which cut our country into two in 1971".[11]

G. Parthasarthy, former Indian Ambassador to Pakistan, holds a similar view: "Pakistan decided to acquire nuclear weapons not because India had nuclear weapons, but because its ruling elite believed that after the dismemberment of the country during the Bangladesh conflict, it needed such weapons because of India's size and conventional superiority."[12]

In this respect, Pakistan adopted a doctrine and strategy not very different from that pursued by the North Atlantic Treaty Organisation (NATO) against the USSR. Pakistan visualised nuclear weapons as the sole guarantor of its national pride and national survival, and, thus, started to seek Chinese assistance for its nuclear weapon programme in the late 1960s in the aftermath of the US sanctions. Nuclear weapons for Pakistan were seen as a means to neutralise Indian conventional military superiority, and also a projection of its scientific and technological capabilities. In addition, it has perceived nuclear weapons capability as a "currency of power" which would enhance its strategic image in the international community in general and in the Muslim countries in particular.

Pakistan's strategy includes its deep desire to emerge as a formidable country of the Muslim world and to be visualised as the *leader of the Islamic world*. Bhutto, in his death cell testimony stressed:

> .......We were on the threshold of full nuclear capability when I left the government to come to this death cell. We know that Israel and South Africa have full nuclear capability. The Christian, Jewish and Hindu civilisations have this capability. The Communist powers also possess it. Only the Islamic civilisation was without it; but that position was about to change.[13]

In its pursuit of nuclear power status, Pakistan received financial support from Saudi Arabia and Libya, and eventually it did share nuclear data and expertise with Iran, Libya and Iraq. Pakistan believed that once it was able to achieve a nuclear weapons capability, the demand for Pakistani skilled manpower and technology would go up in the oil-rich Muslim countries. In return, Pakistan would have easier access to the oil wealth of the West Asian nations to fulfil its needs for technological development.[14] Pakistan's appeal for financial assistance for the acquisition of a nuclear bomb for the "sake of Islam" was a strong emotional appeal to the Muslim world. Pakistan started receiving significant financial support from the Arab world by the mid-1970s. By mid-1976, five Arab countries and Iran had provided grants and loans worth nearly $ 1,000 million. In a matter of three years (between 1973-74 and 1976-77), the Muslim countries ranked first in the list of aid donors to Pakistan.[15]

After Pakistan attained nuclear status, it obviously felt technologically superior and positioned itself as the leader of the Islamic world. The first foreign envoy who arrived in Pakistan, after the nuclear tests in May 1998, was Kamal Kharrazi, the Iranian Foreign Minister. Iran was the first to express total support for Pakistan in the name of Islamic solidarity. "Having lived for long under the shadow of Israel's nuclear threat, Muslims all over the world feel confident because a fellow Islamic nation has crossed the nuclear threshold," he said. Hamas leader Shaykh Ahmad Yassin hailed Pakistan's capability as an "asset to the Arab and Muslim nations."[16] Although many Pakistani leaders have expressed their discomfort with the use of the phrase 'Islamic bomb', the fact remains, that Pakistan's nuclear weapons capability is seen throughout the Muslim world as an Islamic bomb, which would be potentially used in the interest of Islam.

For more than six decades, Pakistan's foreign and security policies have revolved around how to balance and counter the Indian "threat". The managers of Pakistan's security policy have addressed this dilemma by adopting two classical approaches:

(i) Alliances with major powers to acquire high technology defence capability; and (ii), since the 1960s, nuclear deterrence, to offset India's conventional superiority.[17] Pakistan's perceived threat and insecurity motivated it for alliances with the major powers, resulting in its entry into the Western military alliances, SEATO and CENTO, in the 1950s. Pakistan's alliance with the United States has played a major role in its military build-up. Although its military alliance with the US has been shaped solely by the US strategic requirements and decisions, and not essentially by Pakistani aspirations to acquire high technology weapons, Pakistan did receive robust military and economic aid from the US in the 1950s and again in the 1980s. It was not just the US equipment and defence technology which was important for Pakistan but what also added to the qualitative boost was the military training for Pakistani forces in the US military schools. Islamabad received Western intelligence apparatus, radars and communication equipment, which gave it enough confidence to initiate its aggression in 1965. The US arms supply and also the assistance in varied forms came to a halt following the 1965 War. This was the period when the ruling elites in Pakistan seriously started making efforts to diversify the means and sources of military modernisation.

Lt Gen Sardar F.S. Lodhi, clearly stated the need for nuclear weapons in the wake of Pakistan's weak conventional build-up as compared to India:[18]

> Unfortunately, in South Asia, a balance of power cannot be maintained by conventional means alone. Owing primarily to India's sheer size and ample resources. India is larger than all her neighbours combined in South Asia by a wide margin. Add to this India's ambitions across her frontiers in the region and beyond and you have a situation fraught with long-term defence and security implications for Pakistan. India's recent large scale military maneuvers on land and sea, and in the air often close to the Indo-Pak border and her acquisition of 1.6 billion dollars worth of modern arms from Russia is certainly a cause of some concern in Pakistan.

He further added:

> During any future Indo-Pak armed conflict, India's numerical superiority in men and conventional arms is likely to exert pressure beyond endurance. In a deteriorating military situation when an Indian conventional attack is likely to break through our defences or has already breached the main defence line, causing a major setback to the defences, which cannot be restored by conventional means at our disposal, the government would be left with no other option except to use nuclear weapons to stabilise the situation. India's superiority in conventional arms and manpower would have to be offset by nuclear weapons. The political will to use nuclear weapons is essential to prevent a conventional armed conflict, which would later on escalate into a nuclear war.

Nuclear deterrence was emphasised for the first time by Zulfiqar Ali Bhutto in the late 1960s when he clearly stated in his book *The Myth of Independence:*

> All wars in our age have become total wars; all European strategy is based on the concept of total war; and it will have to be assumed that a war waged against Pakistan is capable of becoming a total war. It could be dangerous to

> plan for less and our plans should, thus, include the nuclear deterrent. Difficult though this is to employ, it is vital for Pakistan to give the greatest possible attention to nuclear technology, rather than allow herself to be deceived by an international treaty limiting this deterrent to the present nuclear powers.[19]

Pakistan's perception of threat, which formed the basis of its nuclear will, undoubtedly, continues to dominate its nuclear strategy and force build-up. In the aftermath of the nuclear tests in 1998, Pakistan's primary concern, to maintain a *credible minimum nuclear deterrence* against India's conventional and nuclear threat is unlikely to change in the future.

Pakistan's grand strategy has been centred around hostility against India. The nation has dedicatedly pursued the strategy of covert war through terrorism with gradually intensifying means. The desire to match India's conventional build-up has motivated Pakistan to acquire high technology weaponry and to build up nuclear weapons. Pakistan's military strategy over the last six decades has relied on the following factors:

- Pakistan sought external assistance mainly from the United States and China to build up its military power and to project itself as an ally of a major power in the world scenario. With the suspension of US aid, China emerged as a major support for Pakistan. Pakistan developed close affiliation with China due to mainly two reasons: firstly, China's strategic interest in Pakistan to counter America's growing military alliance and also to balance India's growing superiority. Secondly, the suspended US weapon supplies due to the arms embargo, initially in 1965 and later in the 1990s, motivated Pakistan for diversification of weapon resources. Pakistan provided an easy market for Chinese weapons. Pakistan viewed China as a much more reliable partner than the US. The weapons from China were not only cheap, but also credit was available on easy repayment terms. Chinese assistance came in not only in the form of cheap defence equipment but also for the development of indigenous facilities for defence production in Pakistan. China assisted Pakistan to develop its nuclear capabilities. Apart from these two major powers, Pakistan has been receiving financial, diplomatic and political support from the Muslim world.

- Pakistan has relied more on high technology weapons to seek competitive military advantage.
- Pakistan has believed in offensive aggressive strategies and has had a deep-rooted belief that by taking the initiative and going on the offensive, smaller size forces in history have won wars against bigger enemies. All the four wars which Pakistan has fought with India (in 1947-48, 1965, 1971 and 1999), have been initiated by Pakistan, each with an element of *surprise* and *preemption*. The war in 1971 was caused by Pakistan's internal instability. But the actual war was initiated by Pakistan with a preemptive air strike against Indian Air Force bases on December 3. In addition, it adopted the offensive route for its covert war through terrorism in J&K since 1988 (besides that in Punjab in 1983-93).
- A close study of Pakistan's covert war activities indicates that the factor of religion started to gain momentum in the 1960s in J&K. Religion has been an important factor behind Pakistan's struggle for Kashmir but use of the religion to motivate people to join the struggle and also seek support from the locals in the Valley started in the 1960s. Though it did not provide support to the Pakistani invaders of Force Gibraltar in 1965. Post 1971 War, after the creation of Bangladesh, Pakistan deeply felt the need to use religion as a major component of its grand strategy in order to prevent further disintegration of Pakistan.

Pakistan's Islamisation, especially of its army, had started to grow during Zulfiqar Ali Bhutto's rule. Gen Zia systematically reinforced the Islamisation of Pakistan which was propagated by his predecessor. During Gen Zia's regime, Islamic practices in day to day life were made mandatory by the state. Zia was entrusted with the task of Islamisation of the Pakistan Army when he took over as the Army Chief. For Bhutto, Zia was the right man to take the Pakistani military to the next stage of its evolution as the guarantor of an anti-India, Islamic ideology.[20] Zia, even as a Corps Commander, distributed books written by the Jamaat-e-Islami's founder, Maulana Sayyed Abdul Ala Maududi, as rewards for the officers who proved their calibre in various fields in his garrison. Gen Zia ul- Haq's early steps to Islamise the army are identified by Lt Gen Jahan Dad Khan as Deputy Martial Law Administrator and Corps Commander:

> A devout Muslim, it was a matter of faith with [Zia ul-Haq] to propagate Islam wherever he could. Immediately after his appointment as COAS [Chief of the Army Staff] the motto he gave the troops was *Eman* (Faith), *Taqwa* (abstinence), *Jehad Fi Sabeelillah* (war in the way of or for the sake of God). He urged all ranks of the army during his visits to the troops as well as in written instructions, to offer their prayers, preferably led by the commanders themselves at various levels. Religious education was included in the training program and mosques and prayer halls were organized in all army units.[21]

Zia in his first speech as Chief Martial Law Administrator described himself as a "Soldier of Islam" and declared his commitment to building a new political, economic and social order based on religion.[22] Adherence to Islamic practices emerged as a major criterion for selection in the army. According to Zia, a soldier had to be a true Muslim before taking on the responsibility of defending the Islamic Republic. Pakistan under Zia's regime, started implementing Islamic practices much more as compared to earlier political or military regimes in the country. The state was now being run on the values interpreted on the basis of the Holy Quran.

Religious ideology completely dominated domestic policies in Pakistan in the functioning of the state, besides its foreign and strategic polices. It was during this process of *Islamisation* that the Afghan War came into the picture and altered the total dynamics of Pakistan's political, military and strategic policies.

**Pakistan's Nuclear Doctrine**

In the pre-nuclear test period, Pakistan's doctrine was that of ambiguity. Although, Pakistan even today does not have an officially announced doctrine, statements made by responsible policy-makers in Pakistan have clearly outlined the basic elements in its nuclear doctrine. There is an unofficial code adopted by the Pakistani leadership, based on *Indo-centricity, credible minimum deterrence, strategic restraint* and *first use*. Very interestingly and rather ironically, the code emphasises on the principles of a peaceful programme revolving more around maintaining a balance against the Indian force build-up, but it includes making a first strike in response to not only a conventional attack by India but also a posed threat from India.

***Indo-centric***

Pakistan has built up its defence capability highlighting the perception of a threat from India. Threat perceptions in Pakistan, forming the core of the defence policies, and the military strategy, have remained India-centric. Immediately after partition, Pakistan strongly felt the need to control J&K and incorporate it into Pakistan. Pakistan's initial concerns regarding the annexation of Kashmir were geo-strategic and economic. But this does not rule out the factor of religion which played a major role in Pakistan's persistent claim for Kashmir, in contradiction to Muhammad Ali Jinnah's strong stand that the state ruler's decision regarding accession should be treated as the final criterion for accession in 1947. This eventually shaped Pakistan's grand strategy against India and also India's image as a big power, with hegemonic ambitions against Pakistan.

India remained a primary factor on the military's national security agenda and this has made the military more important than any other player in Pakistan. Fear of India, also became a convincing factor to legitimise the military's financial autonomy and Pakistan's massive investment in the defence build-up which led to a steep rise in the fiscal deficit in the 1990s. The social sector in Pakistan has been deprived due to the diversion of funds towards non-development expenditure. Pakistan has maintained that to counter the Indian threat, it needs to build up the conventional military capability along with the nuclear deterrent. Pakistan's nuclear doctrine and strategy has been formulated solely around the assumed Indian threat of conventional war and its nuclear use doctrine is India specific.

***Minimum Nuclear Deterrence***

This is one of the basic tenets of Pakistan's nuclear doctrine. The concept of credible minimum deterrence is not based specifically on the numbers but the weapon arsenal, including the nuclear weapons, delivery systems, command and control and the doctrine and strategy, based on the perceived threat perception from India. A credible minimum deterrent force means a minimum force capable of inflicting nuclear destruction on India. Prime Minister Nawaz Sharif very distinctly talked about it in May 1999 when he highlighted the key elements of Pakistan's nuclear policy at the National Defence College: "Nuclear restraint, stabilisation and minimum credible deterrence constitute the basic elements of Pakistan's nuclear policy."[23]

Pakistan's Foreign Minister Abdul Sattar repeated the doctrine in November 1999: "Minimum nuclear deterrence will remain the guiding principle of our nuclear strategy."[24]

President Gen Pervez Musharraf also reasserted the concept when he said: "We refuse to enter a nuclear arms race, and seek stability in the region. Pakistan, unlike India, does not have any pretensions to regional and global power status. We are committed to a policy of responsibility and restraint by maintaining a credible minimum deterrent."[25]

A perceptive Pakistani commentator had stated that Pakistan "has to copy the qualities outlined in the Indian doctrine".[26] Therefore, Pakistan has projected its nuclear doctrine as credible minimum deterrence, but one which is India specific and more manageable.[27] The reasons for Pakistan's adoption of credible minimum deterrence are obvious. Pakistan desires a viable nuclear arsenal as the whole logic of going nuclear was Pakistan's inability to cope with India's conventional build-up, primarily due to the financial constraints.

The term 'minimum' begs definition and can be interpreted differently by different states. It is unclear that Pakistan has actually quantified deterrence, and minimum for Pakistan would be based on the calculation of the threat, owing to the numbers of the nuclear weapons and missile systems with India. Pakistan's minimum deterrence appears to be based on the capability to inflict unacceptable damage or assured destruction. Zafar Nawaz Jaspal wrote:

> It seems that in the present strategic scenario, Pakistan possesses enough strategic weaponry capability to provide it with a minimum nuclear deterrence. The basis of this perception is that in nuclear deterrence, parity between opponents is not based on numerical equality of the number of nuclear delivery systems, or of the number of warheads or in the yield of megatons available to each opponent. Parity requires assured destruction capability.[28]

### *First Use Doctrine*

Pakistan has long held the belief that being the weaker state, it can compensate for that weakness by taking a bold initiative, preferably with strategic surprise, to attack the Indian military capability and, thus, reduce the adverse margin of capabilities. This was the military strategy it practised in all the wars it waged against India, including the last one which was a regular war in Kargil in 1999 and, more important, the war through terrorism across the border for a quarter century. The specific concentration of terrorism in the border districts in Punjab west of the river Beas was clearly aimed at similar goals. Seen in the context of this strategic mindset, it is not surprising that it has adopted a nuclear doctrine of "first use". In fact, it has often claimed that it would/could use nuclear weapon at the very beginning of the war with India, if the Indian military even crossed the international border.

Pakistan's first use doctrine has been clearly stated by Lt Gen Sardar F.S. Lodhi (Retd) in the *Pakistan Defence Journal*:

> The political will to use nuclear weapons is essential to prevent a conventional armed conflict, which would later on escalate into a nuclear war.........Pakistan's Nuclear Doctrine would, therefore, essentially revolve around the first-strike option. In other words, we will use nuclear weapons if attacked by India even if the attack is with conventional weapons. With his American experience of a graduated nuclear response Professor Stephen P. Cohen feels that Pakistan would use what he calls an 'option-enhancing policy' for a possible use of nuclear weapons. This would entail a stage-by-stage approach in which the nuclear threat is increased at each step to deter India from attack. The first step could be

> a public or private warning, the second a demonstration explosion of a small nuclear weapon on its own soil, the third step would be the use of a few nuclear weapons on its own soil against Indian attacking forces. The fourth stage would be used against critical but purely military targets in India across the border from Pakistan. Probably in thinly populated areas in the desert or semi-desert, causing least collateral damage. This may prevent Indian retaliation against cities in Pakistan. Some weapon systems would be in reserve for the counter-value role.[29]

Rejecting India's proposal for a joint no first-use pledge in the aftermath of the nuclear tests, Pakistan took the stand that it would be the first to use nuclear weapons to counter India's conventional capability. Lt Gen Sardar Lodhi (Retd) has justified Pakistan's dismissal of India's offer:

> India's offer of a treaty to be signed by the two countries, agreeing not to be the first to use nuclear weapons against each other is one-sided and would benefit India only, as it has a superior conventional force. It may be more apt for both countries to sign a mutual test ban treaty to start with, followed by a no-war pact.[30]

Pakistan's argument has been that in case of the likelihood of a conventional attack, or in a situation when India has breached the defence line, causing a major setback to the defence and security arsenal of Pakistan, then, due to the fear of being defeated in a conventional war against India, Pakistan has to resort to the first use option. Pakistan will go for the nuclear weapons first even if the attack from the Indian side is with the conventional weapons. Thus, a first use policy, according to the Pakistani leadership, provides credible security guarantees to their national sovereignty.

The same logic for first use was used by NATO during the Cold War when it suggested that it would be the first to use nuclear weapons in a conflict as it perceived that a hostile Soviet Union had an overwhelming advantage in conventional forces. Following the collapse of the Soviet Union, NATO has tried to play down the role of nuclear weapons but it maintains the doctrine which gives it the leverage to use nuclear weapons first in case of a conventional attack. Given the limited resources, geo-strategic proximity, India's growing power and Pakistan's concern for strategic depth, the Pakistani leadership's claim has been that it cannot afford to adopt the no-first-use policy. A major factor which has led to Pakistan's first use posture, apart from India's conventional superiority, is the fact that a first strike nuclear force is affordable in financial terms, and less complicated to build.[31]

The need for Pakistan to maintain this position of first use is further exacerbated with India's assertion that the possibility of a conventional confrontation is not ruled out even in the nuclear environment. This obviously creates disturbances within Pakistan because it stands in contrast to Pakistan's rationale for the nuclear weapons which is to deter India from using conventional force against it.[32] To deter India from using conventional force, Pakistan holds the position of first use, in case, it "feels" the need. This came out very openly, much before the Pokhran tests in 1987, when A.Q. Khan in an interview to Kuldip Nayar said: "Mr Nayar, if you ever drive us to the wall, we will use the bomb. You did it to us in East Bengal. We won't waste time with conventional weapons. We will come straight out with it."[33]

### *Possibility of Preemption*

Pakistan's obsessive reliance on the doctrine of first use seems to be emerging from two factors. First, Pakistan wants to keep an option open for "preemptive nuclear strikes" against India and it is convinced that its preemptive strikes would lead to the destruction of India's retaliatory capabilities and/or paralyse the Indian political decision-making. Secondly, Pakistan has failed to consider the consequences of the Indian retaliation. Pakistan seems to assume that India would not use nuclear weapons against it even after getting hit. An assumption related to India's strategic culture could be one of the pressing reasons for the Pakistani assumption which seems to be intensified with India's non-reactive approach (with conventional means) towards the Pakistan sponsored terrorist attacks.

India, in its nuclear strategy and doctrine, has adopted 'restraint' as a responsible and politically mature nation-state. But the Indian restraint cannot be read by Pakistan as an unending and open-ended policy of the Indian national state. India shall resort to retaliation to a Pakistani nuclear strike given the situation, and the consequences for Pakistan would be fatal.

The scenarios in which Pakistan would opt for nuclear weapons have been spelled out by Pakistani elites who have cultivated a posture of irrational rationality to try and enhance the effect of this posture. Lt Gen Khalid Kidwai, the head of the Strategic Planning Division in Pakistan's nuclear command and control system, in an interview (published in 2002) to Italian professors, claimed that "nuclear weapons would be used only if the very existence of Pakistan as a state is at stake". But he went on to state that Pakistan would definitely use nuclear weapons in case deterrence fails and:

- If India attacks Pakistan and conquers a large part of its territory (space threshold);
- If India destroys a large part either of Pakistan's land or air forces (military threshold);
- If India proceeds to the economic strangling of Pakistan (economic strangling);
- If India pushes Pakistan into political destabilisation or creates large-scale internal subversion in Pakistan (domestic destabilisation).[34]

The last two elements, based on economic strangulation and domestic destabilisation, cannot be seen in isolation and it would not be incorrect to say that these vaguely draw vast parameters which can potentially include various steps taken up by the Indian government in the wake of acute tension or conventional confrontation. Peter R. Lavoy has said that these two elements are "offshoots or preludes to a conventional war that India might undertake".[35] Economic strangulation could imply an Indian naval blockade or possibly also the placement of Indian dams on rivers flowing from Kashmir that could lead to a water crisis in terms of either shortage or floods in the Punjab province of Pakistan.[36] The fourth scenario related to domestic destabilisation emerges from Pakistan's perceived threat from India which was deepened after the creation of

Bangladesh in 1971. Pakistan's concern about its national survival is exacerbated due to India's role in Afghanistan and support for the Karzai government, and Pakistan's consistent claim regarding the Indian support to the Baluch insurgency.

Kidwai's outlined scenarios create ample amount of uncertainty by covering a wide range of situations and potential actions by India during a crisis. It is basically to deter India from any action, military, economic or diplomatic, which could hurt Pakistan in any form. Peter Lavoy very rightly noted:

> Kidwai's four existential threats for possible use are credible but also vague. The statement was almost intended to be imprecise so as to enhance Pakistani deterrence. If Pakistan were more explicit about nuclear red lines, this might enable India to adjust the scope of its strategic plans and military operations accordingly. By not specifying the precise Indian actions that would trigger Pakistan's use of nuclear weapons, Pakistan's defence planners hope to create uncertainty in the minds of Indian policymakers as to how far they can press Pakistan on the battlefield.[37]

The objective of Pakistan's nuclear weapons is to deter any form of Indian military response and, thus, Pakistan has deliberately adopted a posture of irrationality. The Pakistani leadership is convinced that they have managed to deter India with their posture of irrationality and uncertainty, more than once. This interpretation obviously came out very clearly in President's Musharraf's December 2002 statement. Although Musharraf did not specify the nuclear threat in his speech to the army corps union in Karachi, he said that he was ready to take a decision and act during the 2002 crisis:

> In my meetings with various world leaders, I conveyed my personal message to Indian Prime Minister Vajpayee that the moment Indian forces cross the Line of Control and the international border, they should not expect a conventional war from Pakistan. I believe my message was effectively conveyed to Mr. Vajpayee.[38]

***Weapon of Last Resort***

Most of the Pakistani writings pre-1998 pointed towards build-up of the nuclear capability against the Indian conventional forces and, thus, implied first use. But there seems to be a shift in the Pakistani thinking regarding the use of nuclear weapons and adoption of a relatively moderate stand by claiming nuclear weapons as the "weapons of last resort".

President Pervez Musharraf, in May 2002, said that if it came to war between the nuclear-armed rivals, he would "respond with full might." These statements were interpreted to mean that if pressed by an overwhelming conventional attack from India, which has superior conventional forces, Pakistan might use its nuclear weapons. A similar interpretation can be made from the statement of Pakistan's Defence Minister Rao Sikander: "The country's ultimate security lies in the use of the atom bomb; it is not a mere showpiece."

Pakistan perceives nuclear weapons as the ultimate guarantee of its security and national survival but has not been very clear in defining the red lines and the scenarios which could lead to its first use. Abdul Sattar (former Pakistani Foreign Minister), Agha Shahi and Air Chief Mshl Zulfiqar Ali Khan jointly authored an article in *The News* on October 5, 1999, which stated:[39]

> The exigency under which the Pakistan Army may use nuclear weapons is spelt out as: although the precise contingencies in which Pakistan may use nuclear weapons have not been articulated or even defined by the government, the assumption has been that if the enemy launches a war and undertakes a piercing attack to occupy large territories or communications junctions, the *weapon of last resort* would have to be invoked.

In April 2002, in an interview published in the German magazine, *Der Spiegel*, Musharraf said: " If the pressure on Pakistan becomes too great then as a last resort, the [use of] atom bomb is also possible."[40]

Musharraf's statement of last resort was made in 2002 and in the same period, in his address to the army corps union in Karachi, he said that war with India was averted due to his repeated warnings of using "unconventional" means (interpreted mostly erroneously as nuclear weapons) in the case of India breaching the red lines. This no doubt (also) implied use of a large number of guerrilla/*jihadi* fighters rather than nuclear weapons. Otherwise, there is a contradiction in Pakistan's stance where, on the one side, it claims to use its nuclear weapons as a last resort, and on the other, is convinced that the threat of use of nuclear weapons was successful in deterring the Indian military posture.

A weapon of last resort can be logically interpreted as a scenario wherein no other means are left with a nation to defend itself. In the case of a conventional war between India and Pakistan, if Pakistan's military reserves are destroyed, it may not have the capacity to defend itself and rather than losing the war, it would opt to use nuclear weapons against India. Pakistan's use of the nuclear weapon, even as a last resort, would face Indian retaliation, which talks about "unacceptable damage". Although, on the one hand, the weapon of last resort option stands in contradiction to Pakistan's earlier statements, and appears moderate, on the other, it also projects a mindset for self-destruction, where complete destruction of the nation is preferred over all other possible options.

Although there has been no official statement made by Pakistan quantifying its threshold, some analysts believe that if Indian forces crossed the N-5 highway connecting Lahore to Karachi, Pakistan might escalate a conventional conflict into a nuclear one.[41]

***Nuclear Policy Based on Restraint and Responsibility***

In the last few years, Pakistan's endeavour has been to project itself as a *responsible* nuclear power. The need to do so was exacerbated with India and the US signing the nuclear deal. Pakistan has been keen for a similar nuclear agreement with the US and, thus, projection of a responsible nuclear posture became inevitable. In 2006, Lt Gen. Khalid Kidwai, in his address to the Naval Postgraduate School, Monterey, said that Pakistan has dealt with formidable challenges by developing a nuclear policy based on "restraint and responsibility" with four salient features: "(1) deterrence of all forms of external aggression; (2) ability to deter a counter-strike against strategic assets; (3) stabilisation of strategic deterrence in South Asia; and (4) conventional and strategic deterrence methods."

Again, Gen Khalid Kidwai talks about "deterrence to all forms of external aggression", which should include both conventional and nuclear aggression. This is in line with the statements made by the Pakistani policy-makers in the past.

Deterring India's conventional posture remains the prime objective of Pakistan's nuclear weapons. Kidwai talks about building the ability to deter counter-strikes against strategic assets. Pakistan has expanded its arsenal and delivery systems substantively in the last ten years, in order to threaten a disarming strike to wipe out or at least drastically reduce, India's retaliatory capability [deter a counter-strike].

Thus, looking into Pakistan's stance on nuclear weapons, numerous contradictions can be seen. It talks about use of nuclear weapons as a last resort, with restraint and responsibility. Obviously, last resort has not been defined by Pakistan and, thus, creates uncertainty, which enhances deterrence and serves the Pakistani objectives. It would not be incorrect to state that Pakistan has very rationally adopted the posture of irrationality.

## Pakistan's Nuclear Strategy

### *Pakistan's Covert War and Strategy of "Offensive Defence"*

Pakistan has tried to project its nuclear assets as an instrument of blackmail. The acquisition of nuclear capability enhanced Pakistan's capability to wage and escalate the covert war in Kashmir. Professor Stephen Cohen, during his visit to Pakistan in 1980, in his discussions with various defence personnel, carried the impression that according to many Pakistanis nuclear capability:

> Would neutralize an assumed Indian force. Others point out, however, that it would provide the umbrella under which Pakistan could reopen the Kashmir issue; a Pakistani nuclear capability paralyses not only the Indian nuclear decision but also Indian conventional forces and a brash, bold, Pakistani strike in Kashmir might go unchallenged if the Indian leadership was weak or indecisive.[42]

Pakistan's non-adherence to no first use was believed to serve the objective of deterring India from responding with conventional military retaliation. Policy-makers in Pakistan seem to be convinced that they will be able to carry on, or rather accelerate, their activities in Kashmir under the broader threat of using nuclear weapons, if required, and this would constrain India's strategic moves. Although this has been the Pakistani thinking for long, it has grown tremendously with Pakistan's acquisition of nuclear weapons and announcement of the first use policy.

Pakistan claimed that it had the capability to build a nuclear bomb in 1987, and it is noteworthy that during the late 1980s, the activities in the Valley also witnessed a shift. The terrorist acts increased significantly in numbers and were planned in a more organised manner. Pakistan became more vocal about the possession of nuclear weapons in the late 1980s, and then in the 1990s, in order to create an impression that any radical move from the Indian side might lead to retaliation with the *Islamic bomb.*

Pakistan's grand strategy has been to use the nuclear card for aggressive militancy in Kashmir in the 1990s and eventually terrorism in other parts of Indian territory. In 1989, the then Army Chief, Gen Aslam Beg asserted that Pakistan lost the previous wars with India due to "lack of a clear strategic vision." He announced that Pakistan had a coherent strategy now. "One aspect of the strategy was launching of the militant proxy war in J&K from July 31, 1988; and the other was the achievement of nuclear deterrence to provide 'defence' in the offensive-defence strategic doctrine, while irregular war was used for the 'offensive' component by 1987".[43] Aslam Beg also announced that the next war for Pakistan would be in the 'enemy's territory', which indicated the escalation of terrorist activities in the Indian territory supported by Islamabad.

At the peak of *jihadi* terrorism in Kashmir, on August 23, 1994, Nawaz Sharif, the former Prime Minister of Pakistan, during his visit to Kashmir, declared that Pakistan was a nuclear power. "I confirm Pakistan possesses the atomic bomb", he announced and then warned India that an attack on Pakistan could trigger a nuclear war.[44] Pakistan's policy of first use was obviously used as a strong card for its covert activities not only in the Valley but also other parts of India where Pak-sponsored terrorism has accelerated over the years. The policy of first use, in the Pakistani thinking, serves as a deterrent and restrains its neighbour from massive military retaliation. Air Cmde Jasjit Singh has very rightly stated that the "very rationale of going nuclear was to neutralise India's vastly superior conventional military potential. Under the nuclear umbrella, Pakistan has felt confident of prosecuting its low-cost proxy war against India".[45]

The JIC (Joint Intelligence Committee—India) emphasised, in August 1991, how Pakistan might use its nuclear capability to advance its objectives in Kashmir.[46] "Pakistan's nuclear capability would allow Pakistan to continue its Low Intensity Conflict (LIC) in Jammu and Kashmir and Punjab under the cover of a nuclear threat, which in their perception would limit India's retaliatory options of threatening escalation to a conventional war."

Pakistan used nuclear weapons not only to blackmail India but also to send a message to the international community so that it would pressurise India not to retaliate militarily in the stressed situations and help Pakistan to attain its diplomatic and political objectives. In July 1995, the JIC Monthly Intelligence Review (MIR) assessed that "statements made by Pakistan Foreign Ministry officials attempt to use the nuclear issue to attract international attention to the Kashmir issue for their diplomatic ends."[47]

As early as in 1987, when Pakistan acquired the nuclear weapon capability, it conveyed a nuclear threat to India at the time of Operation Brasstacks. To diffuse the Brasstacks crisis, Pakistan relied on its nuclear weapons capability. According to the Kargil Committee Report, "This was officially communicated by Pakistan's Minister of State for Foreign Affairs, Zain Noorani, to the Indian Ambassador in Islamabad. It was also communicated by the Pakistani nuclear scientist, Dr A.Q. Khan to the Indian journalist Kuldip Nayar."[48]

Pakistan did not directly communicate the nuclear threat to New Delhi through the official channels, but instead facilitated a prominent Indian journalist, Mr Kuldip Nayar to interview Dr A.Q. Khan, who distinctly stated that Pakistan possesses nuclear weapons: "America knows it. What the CIA has been saying about our possessing the bomb is correct and so is the speculation of some foreign newspapers".[49] The interview was deliberately arranged to warn the Indian leadership about the consequences of militarily attacking Pakistan.

In August 1990, Indian intelligence obtained information that Pakistan had developed a policy of using nuclear weapons as a first resort in the case of a war.[50] Some Western reports suggested that Pakistan had manoeuvred its nuclear capability.[51] The idea that nuclear weapons have served to deter India on "at least three occasions" was articulated by former Pakistan President Ghulam Ishaq Khan in July 1993, and in 1999, in a joint article by Pakistan's Foreign Minister, Abdul Sattar, Aga Shahi and Air Chief Mshl Zulfiqar Ali Khan.[52] Pakistan's belief that India would be deterred by Pakistan's 'first use' doctrine formed one of the most critical bases for the Kargil intrusions. The Kargil Review Committee Report stated regarding the assumptions of Pakistan's aggression in Kargil:

> Its nuclear capability would forestall a major Indian move, particularly across the international border involving use of India's larger conventional capabilities. It appears to have persuaded itself that nuclear deterrence had worked in its favour from the mid-1980s.

During the Kargil War, Pakistan did threaten the use of nuclear weapons in case of escalation of the conflict from the India side. Post Kargil, in 2001, following the attack on the Indian Parliament on December 13, 2001, India mobilised its forces and Operation Parakaram was conducted. Pakistan once again used the nuclear card in its efforts to build international pressure on India as the 2001-02 military stand-off between the two nuclear power states was indeed alarming. President Musharraf in his speech on March 23, 2002 said: "With the blessings of Allah, Pakistan has the full capability to defend itself and is in a position to thwart any aggression against it. If anyone tries to challenge its power, they would be taught a lesson which would be remembered."[53]

In a subsequent address to the nation on May 27, 2002, President Musharraf announced:[54] "We do not want war. But if war is thrust upon us, we would respond with full might, and give a befitting reply".

Although the term nuclear has been deliberately avoided, the military leadership did threaten the use of nuclear weapons which would inflict "unforgettable damage". The Pakistani leadership has been consistent in using the nuclear threat to counter Indian conventional retaliation, directly and indirectly, by building up international pressure on India to refrain from military moves.

### *Nuclear Weapon as a Political Tool*

Although the build-up of its nuclear arsenal was to counter-balance India's conventional superiority, Pakistan has used its nuclear capability to fulfil its political and diplomatic objectives. Pakistan today stands as the epitome of terrorism, rising *jihadism* and escalating Talibanisation, along with its depressing economy. What makes the threats emanating from Pakistan more worrisome is the fact that Pakistan possesses nuclear weapons and may not have clearly absorbed the full implications of their use. The fear of nuclear weapons falling into the hands of non-state actors or even radical *jihadi* elements within the military has compelled the United States to surrender to the demands of Pakistan. Although one can argue that Pakistan has no option but to align with the US, despite the domestic resistance, the fact is that the US today definitely needs Pakistan to fight its war on terror in Afghanistan and also, Pakistan's nuclear weapons have placed the superpower in a vulnerable position, making it incapable of taking aggressive action against the state. The last eight years have witnessed a rapid growth in the American weapon supply and financial aid to Pakistan, despite Washington giving clear indications of dissatisfaction with the Pakistan military's performance in the war on terror. The United States has not succeeded in getting Pakistan to alter its strategy of supporting extremist elements despite strong warnings which come along with the lavish US military and economic assistance.

### *Nuclear Weapon as a Cover for Terror*

Pakistan has used the nuclear card successfully to continue its covert war and acts of terrorism in India. Over the decades, the military has relied on the strategy of using *terror* as an instrument, which has been supported by the interpretations of the teachings of the Holy Quran.

The most comprehensive and precise study on the war doctrine and strategy in the context of the Holy Quran has been by Brig S.K. Malik in his book, *The Quranic Concept of War*. Brig Malik interprets that "the Holy Quran has given a comprehensive treatment to its concept of war" and "determines all aspects of the use of 'force' in inter-state relations."[55] He regards the Quranic philosophy of war as supreme, which provides the directions for the initiation, planning and control of war. According to his interpretation, the very "initiation of war is for the cause of God" and *jihad* is the "most glorious word in the vocabulary of Islam".

*Terror*, according to the author and a majority of the Pakistani military officers, is central to the war strategy. Use of terror as an instrument to impose one's will and decisions on the enemy has been legitimised, citing examples from the Holy Quran and arriving at the conclusion that "when God wishes to impose His will upon his enemies, He chooses to do so by casting terror in their hearts".[56]

The Islamic concept, as derived from the Holy Quran by Brig Malik, relies on the use of terror in the preparation stages of war itself in order to assure victory and achieve direct results.[57] Thus, terror has been legitimised in the Islamic ideology which has till today shaped the military strategy of Pakistan. Brig Malik asserts, "Terror struck into the hearts of the enemies is not only a means, it is the end in itself."[58] War strategy, according to the Holy Quran, might have different forms depending on various factors but it must be effective and capable of striking terror into the hearts of the enemies. It is emphasised that *fear* has to be an important part of the preparation of war and it determines that "the test of utmost preparation lies in our capability to instill terror into the hearts of the enemies."[59]

The Quranic misinterpretation of the conduct of modern war legitimises *jihad* and the use of terror. One of the most effective modes of generating terror is through *covert warfare* and, thus, creating physical and mental unrest and fear in the enemy territory. Covert warfare, as practised by Pakistan, which includes different tactics like guerrilla warfare, inciting insurgencies and destruction of buildings, works towards generating terror and insecurity in the minds of Indians and, thus, aims to weaken their physical and mental capabilities.

***Conventional Military Build-up***

Pakistan has been focussed on strengthening its conventional military capabilities to deal with limited war below the nuclear threshold plus greater freedom/confidence for covert war. A strong conventional military base definitely provides confidence to continue terrorism, as the nation is strong enough to face Indian retaliation. The US and China have been the main players in building Pakistan's defence capability.

Pakistan has been on an arms acquisition spree to rapidly modernise its air force since the Kargil War, accelerating the process during the last seven years. Significantly, Pakistan's arms acquisitions since the Kargil War have been exclusively centred on modernisation and build-up of the air force and the aerial maritime strike capabilities of the navy. This has to be viewed in the context of the reality that the army in Pakistan, which has ruled the country for most of its existence, and has been in direct control since 1999, calls the shots in military priorities and modernisation. Pakistan has acquired some land systems, but they have been restricted to heavy artillery. Post 9/11, Pakistan has not only received a lavish supply of the US equipment, including the long desired F-16s, P-3Cs, Harpoons and air surveillance radars, but has also managed to continue and accelerate the supply of Chinese defence equipment. Pakistan is jointly producing the JF-17s and (scheduled) J-10s with China. The latest supply from China also includes the Chinese AWACS ZDK-03.

**Targeting Philosophy and Strategy**

Pakistan has consistently displayed an irrational nuclear posture and has created tremendous uncertainty regarding the use of its nuclear weapons. Its targeting the philosophy and strategy would run on the principle of *first use*. The targeting philosophy would not be different from how Pakistan had used its air force for preemptive strikes in the war in 1965 and also in 1971.

During the 1965 War, Pakistan's Grand Slam was an armoured offensive with the active support of the Pakistan Air Force (PAF). Grand Slam was launched on September 1, 1965, in the Chhamb sector to capture the Akhnur bridge. During the 1971 War, the PAF's operational concept was to begin the war with preemptive strikes on Indian forward air bases and radar facilities on December 3, 1971. The overriding priority was to give maximum support to Gen Tikka Khan's proposed coordinated offensive into India with two strike corps, although the grand planned offensive by Tikka Khan's "strike force" did not take place. Again during the Kargil War in 1999, the Pakistan Army had achieved total tactical and strategic surprise. Thus, Pakistan's objective would be to go for preemptive strikes to reduce India's capability.

Secondly, Pakistan has long believed that India finds it difficult to take hard decisions. Hence, it is possible that Pakistan's first strike may target some of India's major cities or at least project a threat to target large cities so as to reduce the possibility of a nuclear exchange.

Taking a hypothetical scenario when Pakistan actually decides to use nuclear weapons against India, in retaliation to India's conventional military response to a terrorist attack conducted by Pakistan-based terrorist group/ groups, its prime objectives would be to two-fold:

Firstly, to reduce India's retaliatory capability.

Secondly, to build-up international pressure on India not to react. The Pakistani elite seem to be convinced that the international pressure will be far too much for India to be able to retaliate.

Given these options, Pakistan would have at least three options in its targeting strategy:

Option I: Target India's nuclear arsenal, in specific, target the delivery systems of the Indian arsenal like the air bases and ballistic missile units.

Option II: Target one or more large Indian cities which would create enough destruction to "paralyse" the Indian command and control system and leadership.

Option III: Try and neutralise the satellite-based communication system to disrupt and possibly black-out the linkages between the National Command Authority, Strategic Forces Command and the launch capabilities. This may require an Anti-Satellite (ASAT) capability.[60] China has already demonstrated such capability in 2007 and it is not unreasonable to assume that it may be transferred to Pakistan. In this context, it should be noted that China has already constructed a satellite for Pakistan and launched it on one of its launch vehicles.

## Conclusion

The central strategic assumption of the Pakistani nuclear strategy has been that India cannot impose a conventional war on Pakistan, leave alone achieve a decisive victory without the risk of the catastrophic consequence of a Pakistani nuclear attack. Pakistan's leadership believes that Pakistan's possession of the nuclear arsenal and its first use policy would be sufficient to deter war. This logic in Pakistan has been reinforced by the common Pakistani perception that the country was able to deter Indian military action on various occasions after highly provocative terrorist attacks, including the one in 2008. Despite the fact that war was narrowly averted in 2002, the Pakistani leadership seems to be very confident of its strategy against India.

This element in Pakistan's grand strategy has encouraged it to enhance its capability to wage and escalate the covert war in Kashmir, as nuclear weapons are believed to deter India from responding with conventional military retaliation. Policy-makers in Pakistan seem to be convinced that they will be able to carry on, or rather accelerate, their activities in Kashmir and the rest of India (like the terrorist strike at Mumbai on 26/11) with the threat of use of nuclear weapons if required and this would control India's strategic moves in the Valley. This process seems to have grown with Pakistan's "first use policy".

In the recent times, there have been signs that the Pakistani elites are becoming increasingly conscious of major limitations of their doctrine and strategy, since Indian strategic experts are talking of being able to exploit the strategic space above terrorism but below the nuclear threshold to enable punitive force being applied to change the covert war strategies and policy in Islamabad which would make its nuclear weapons capability unusable without horrendous costs to itself. Only the future holds the key to what may happen. But the Cold Start doctrine and deployments and the concept of limited war have raised serious doubts about the viability of Pakistan's cherished strategy of covert war and nuclear strikes.

## Notes

1. Ross Masood Hussain, "Threat Perception and Military Planning in Pakistan: The Impact of Technology, Doctrine and Arms Control," in Eric Arnett, ed., *Threat Perception in Pakistan* (Oxford: Oxford University Press, 1997), p. 130.
2. See Ibid.
3. Ibid., p. 131.
4. See Sobia Nisar, "The Pakistan Air Force", at, http://www.defencejournal.com/2001/august/airforce.htm
5. Maj Gen Akbar Khan, *Raiders in Kashmir* (Delhi: Army Publishers), p. 4.
6. Air Cmde Jasjit Singh, *Kargil 1999: Pakistan's Fourth War for Kashmir* (New Delhi, Knowledge World, 1999), p. 2.
7. Zulfiqar Ali Bhutto, *The Myth of Independence* (London: Oxford University Press, 1969), p. 152.
8. Praveen Swami, *India, Pakistan and the Secret Jihad* (New York: Routledge, 2007), p. 148.
9. Ahmed Rashid, *Descent Into Chaos* (London: Penguin, 2008), p. 111.
10. See Hasan Askari Rizvi, *Military, State and Society in Pakistan* (London: Macmillan, 2000).
11. Former Foreign Minister of Pakistan Agha Shahi in Fasahat H. Sayed, ed., *Nuclear Disarmament and Conventional Arms Control Including Light Weapons* (Rawalpindi: Friends, 1997), p. 421.
12. G. Parthasarthy, "The Nuclear Doctrine—For Ensuring Global Strategic Stability", *Business Line*, September 22, 2005.
13. *Transcript of Testimony Before the Supreme Court of Pakistan*, in reply to Pakistan Government White Papers, as cited in *From Surprise to Reckoning: The Kargil Review Committee Report* (New Delhi: Sage, 1999), p. 185.
14. Maj Gen D.K. Palit and P.K.S. Namboodari, *Pakistan's Islamic Bomb* (New Delhi: Vikas Publishing, 1979), p. 17.
15. Ibid., p. 36.
16. Ibn Warraq, "The Islamic Bomb", at, http://www.secularhumanism.org/library/shb/warraq_14_3.html
17. Rahul Bakhsh Rais, "Conceptualizing Nuclear Deterrence: Pakistan's Posture", in E. Sridharan, ed., *The India-Pakistan Nuclear Relationship: Theories of Deterrence and International Relations,* (London: Routledge, 2007), p. 53.
18. Lt Gen F.S. Lodhi, "Pakistan's Nuclear Doctrine", *Defence Journal,* at http://www.defencejournal.com/apr99/pak-nuclear-doctrine.htm
19. Bhutto, n. 7, p. 153.
20. Husain Haqqani, *Pakistan: Between Mosque and Military* (Lahore: Vanguard Books, 2005), p. 112.

21. Lt Gen Jahan Dad Khan, *Pakistan: Leadership Challenges* (Karachi: Oxford University Press, 1999), p. 158, as cited in Haqanni, Ibid., p. 113.
22. Ibid., p. 127.
23. "Remarks of the Prime Minister of Pakistan, Nawaz Sharif, on Nuclear Policies and the CTBT", National Defence College, Islamabad, May 20, 1999.
24. Foreign Minister Abdus Sattar, Address to the "Pakistan Response to the Indian Nuclear Doctrine" Seminar, November 25, 1999, in "Pakistan Responds to India's Nuclear Doctrine", *Disarmament Diplomacy*, Issue No.14, November 1999.
25. http://belfercenter.ksg.harvard.edu/files/uploads/Sagan_MTA_Talk_050708.pdf
26. M.B. Naqvi, *Defence Journal* (Karachi) February 2000, p.15, as cited in Jasjit Singh, " Pakistan's Nuclear Strategy: An Assessment", *Aakrosh* , Vol. 4, No. 13, October 2001, p. 84.
27. Singh, Ibid., p. 84.
28. Zafar Nawaz Jaspal, "Reassessing Pakistan's Nuclear Strategy", *Defence Journal,* at http://www.defencejournal.com/2001/july/reassessing.htm
29. Lodhi, n. 18.
30. Ibid.
31. Bhumika Chakma, *Pakistan's Nuclear Weapons* (London: Routledge, 2009), p. 52.
32. Manpreet Sethi, *Nuclear Strategy: India's March Towards Credible Deterrence* (New Delhi: Knowledge World, 2009), p. 61.
33. *The Telegraph*, November 4, 2007, http://www.telegraphindia.com/1071104/asp/7days/story_8508991.asp
34. As cited in Paolo Cotta-Ramusino and Maurizio Martellini, *Nuclear Safety, Nuclear Stability and Nuclear Strategy in Pakistan,* Landau Network-Centro Volta Report, January, 2002.
35. Peter R. Lavoy, "Pakistan's Nuclear Posture: Implications of Indo-US Cooperation", *Air Power Journal* , Vol. 2, No. 4, Winter 2007, October-December, pp. 147-174.
36. Ibid.
37. Peter R. Lavoy, "Islamabad's Nuclear Posture: Its Premises and Implementation", in Henry D. Sokolski, ed., *Pakistan's Nuclear Future: Worries Beyond War* (Strategic Studies Institute, 2008), p. 137.
38. Cited in Ibid.
39. Agha Shahi, Zulfiqar Khan and Abdul Sattar, "Securing Nuclear Peace", *The News* (Islamabad), October 5, 1999.
40. Cited in H.S. Rao, "Pak May Use Nukes in the Event of War: Musharraf", rediff.com, April6, 2002, at http://www.rediff.com/news/2002/apr/06pak2htm
41. *Jane's World Armies, Pakistan*, Jane's Information Group, 2009.
42. "Nuclear Security and Security Policy in Pakistan", Paper presented at the annual meeting of the Association for Asian Studies, Washington, March 1980, as cited in *Kargil Review Committee Report*, n. 13, p. 187.

43. Gen Mirza Aslam Beg, *The Nation*, December 13, 1993, as cited in, "The Army in Pakistan", in Singh, ed., n. 6, p. 47.
44. Youssef Bodansky, "Pakistan's Nuclear Brinkmanship", *Kashmir Information Network*, at http://www.kashmir-information.com/bodansky/article3.html
45. "The Army in Pakistan", in Singh, ed., n. 6, p. 49.
46. *The Kargil Review Committee Report,* n. 13, p. 197.
47. Ibid., p. 198.
48. Ibid., p. 240.
49. Kuldip Nayar, "We Have the Bomb, Says Pakistan's Dr. Strangelove", *The Observer,* (London), March 1, 1987.
50. ROD with I.K. Gujral (September 24, 1999), and ROD with B.G. Deshmukh (October 26, 1999) as cited in *The Kargil Review Committee Report*, n. 13, p. 66.
51. See James Adams, "Pakistan's 'Nuclear War' Threat", *Sunday Times,* May 27, 1990.
52. n. 37.
53. President Gen Pervez Musharraf's Address to the Nation, March 23, 2002, at http://presidentmusharraf.wordpress.com/2006/07/17/musharraf-address-23-march-2002/
54. President Gen Pervez Musharraf's Address to the Nation, May 27, 2002, at http://presidentmusharraf.wordpress.com/2006/07/20/musharraf-address-27-may-2002/
55. Brig S.K. Malik, *The Quranic Concept of War* (New Delhi: Himalayan Books, 1986), p. 1.
56. Ibid., p. 57.
57. Ibid.
58. Ibid., p. 59.
59. Ibid., p. 144.
60. At present, Pakistan does not have the capability for Ballistic Missile Defences (BMD). Theoretically, given the nature of the China-Pakistan strategic nexus, it will remain a possibility that the Chinese BMD could cover Pakistan's territory, especially the more critical assests. This will have its own impact on the Indian nuclear deterrent.

# 6. Nuclear Command and Control: Safety and Security of the Nuclear Arsenal

Pakistan's nuclear arsenal has undergone a significant expansion in the last ten years. Pakistan has reportedly been building close to 12-15 weapons per year, nearly doubling its warheads in the last three years. Presently, the total number of warheads with Pakistan stands between 90-120.[1] The fissile material production is being expanded through two new reactors. Also, the growth rate of the delivery systems, including the ballistic missiles, aircraft and cruise missiles has been brisk.

The challenge is obviously to guard the nuclear arsenal, including more than 100 weapons and a variety of delivery systems of diverse ranges. The pace at which Pakistan is expanding its arsenal adds to the challenge of securing these weapons. Although, the Pakistani leadership has repeatedly expressed assurance in the security of their nuclear weapons, the United States has expressed its distress rather frequently. In fact, not only abroad but also domestically, there has been enough anxiety regarding the safety of the nuclear weapons.

The safety and security of the nuclear arsenal becomes extremely critical in the case of Pakistan mainly due to three factors.

First, the clandestine nature of the nuclear programme and highly controversial proliferation record, which brings out the loopholes in the political and administrative system.

Second, Pakistan's unwritten doctrine believes in first use and there has very often been talk about using nuclear weapons in the initial stages of a war or even merely in the case of the threat of a conventional war with India. This enhances the probability of accidental use of weapons or use even through miscalculation.

Third, incompetence and corruption is ingrained in the functioning of the Pakistani state system. There is clearly a lack of systematic and structured functioning within the state system where even the simplest affairs of the state are mismanaged, and, hence, the safety of the nuclear arsenal does become questionable. As rightly observed by a Pakistani commentator, "We are unable to efficiently manage simple functions of state, the haphazard traffic on our roads is a reflection of our organisation, how can one expect that the security of our nuclear assets is foolproof?" [2]

Fourth, escalating terrorism within Pakistan, which has expanded to the interiors of Punjab and Sindh, posing a major challenge to the security forces and establishment. The anti-state terrorist organisations like the Tehrik-e-Taliban Pakistan (TTP) (allied with Al Qaeda), high levels of radicalism in the Pakistani society, the insurgencies in Pakistan's periphery and undeniable extremist elements within the Pakistan Army have increased the concerns regarding the safety of the nuclear arsenal in the last 7-8 years. This is greatly bothering the international community, specifically the United States, and according to senior Administration officials, the United States has spent more than $100 million to assist Islamabad to secure its nuclear weapons. Undoubtedly, Barack Obama's worst security nightmare is the possibility of Pakistan's nuclear weapons falling into the hands of Al Qaeda or other non-state actors.

With rising terrorism in Pakistan and frequent attacks on the high security and heavily fortified complexes, the possibility of non-state actors breaking into the nuclear arsenal with insider support is high. In an attack on the Mehran Naval Base in May 2011, the role of insiders was well demonstrated. Under these circumstances, two aspects need attention: one, the nuclear command and control organisation for employment of the nuclear arsenal; and two, the safety and security of the nuclear arsenal and its components.

**Structure of Nuclear Command and Control**

The Pakistani leadership started to take some structured measures towards command and control after the nuclear tests in 1998. Not much is known in the public domain regarding Pakistan's nuclear command and control in the initial phases of the nuclear programme. The civilian and military leadership was primarily focussed on building the weapon and, thus, command and control was not given much attention. The nuclear weapon programme was solely controlled by a few and Dr A.Q. Khan was directly reporting to the Prime Minister's office in the 1970s. A. Q. Khan was initially under the Pakistan Atomic Energy Commission but later, in 1976, Prime Minister Zulfiqar Ali Bhutto provided him autonomous control of the uranium enrichment project and Khan became directly answerable to the Prime Minister himself.[3] The Khan Laboratories were officially given a free hand in pursuing research.

Some of the authoritative sources indicate that a nuclear decision-making committee existed since 1975. One source is Mushahid Hussain, a Senator close to Musharraf, who stated that from 1975 to 1991, the nuclear programme had been controlled by a committee under the supervision of a body headed by Ghulam Ishaq Khan who became the President in 1988.[4] There is not much clarity regarding the nature of this committee. It could be the body created in 1976 to monitor the activities of the Kahuta Research Laboratories (KRL) and the Pakistan Atomic Energy Commission (PAEC) or it could be the troika to oversee the entire nuclear programme: the President, the Prime Minister and the Army Chief.[5] This, in practice, implied that essentially Ghulam Ishaq Khan, Gen Aslam Beg and Benazir Bhutto were tracking the nuclear activities in the later 1980s. Khan shared a close relationship with A. Q. Khan and Ghulam Ishaq, and according to him, he guarded the nuclear programme "like a rock."[6] But undoubtedly, the army remained the most powerful institution, fully dominating the nuclear command and control, leave alone the total power structure of Pakistan. Bhutto commanded little influence in the nuclear decision-making.[7] Benazir had, in fact, 'famously' complained that she was never allowed to visit the nuclear facility. However, senior military officers serving during Benazir's period have refuted this allegation and stated that she never expressed a desire to visit the nuclear facility.[8] The fact is that lack of a clear formal command and control structure controlled by the army benefitted Khan and provided him a free hand with the programme.[9]

In 1996, Gen Jehangir Karamat established a think-tank in the General Headquarters (GHQ), working directly for the Army Chief's Secretariat, basically to provide analytical inputs and conduct studies in strategic affairs.[10] In early 1998, there were discussions regarding some arrangement on command and control, and former Army Chief Gen Aslam Beg did talk about some arrangement which was more of a nuclear decision-making body comprising the Army Chief, the President and senior scientists. But this arrangement was not meant for the operational aspect. According to Brig Naeem Salik, serious efforts in the direction of streamlined command

and control were initiated in 1998 before the tests.[11] After the nuclear tests in 1998, the then Army Chief Gen Jehangir Karamat appointed Maj Gen Khalid Kidwai to oversee an Evaluation and Research Cell (E&R). The recommendations of the E&R cell would, by and large, form the basic guidelines for the command and control system. The E&R recommendations were complete by the fall of 1998 but the unexpected resignation of Gen Karamat and the appointment of Gen Musharraf as the new chief delayed the proceedings.[12] In 1999, the command and control structure was formalised and approved by Gen Musharraf who was prompt in getting the approval of the Cabinet and the National Security Council in February 2000 after the military coup in Pakistan. Khalid Kidwai was appointed as the Director General of the Strategic Plans Division (SPD)—the most critical tier of command and control assigned to formulate the nuclear doctrine and strategy and development of the nuclear arsenal, and he has retained the position even after he retired from the army in 2007.

Pakistan claims that its nuclear weapons are completely secure and its command and control over its nuclear weapons is compartmentalised and includes strict operational security. Pakistan's command and control system is based on "C4I2SR" (Command, Control, Communication, Computers, Intelligence, Information, Surveillance and Reconnaissance).[13] The system is a three-tiered one consisting of the National Command Authority (NCA), the Strategic Plans Division (SPD) and the Strategic Forces Command (SFC).

***National Command Authority***

The NCA, which is the first tier in command and control, was created in 2000. It supervises the functions and administration of all the organisations within Pakistan working on the research, development and employment of nuclear weapons and the military Services that operate the strategic forces.[14] The NCA is the highest decision-making body in Pakistan's nuclear command and control infrastructure. It is a ten-member body and has a mix of civilians and military representatives. The membership includes the Prime Minister as the Chairman, the Ministers of Defence, Interior and Finance, the Director General of the SPD, the commanders of the army, navy and air force, and the Chairman of the Joint Chiefs of Staff.[15] The final authority to take the decision for a nuclear strike requires consensus within the NCA, and the Chairperson, the Prime Minister of Pakistan, casts the final vote.[16]

The NCA is responsible for policy formulation, development and employment of control over all strategic nuclear forces, negotiating arms control/disarmament, overseeing implementation of export controls and safeguarding nuclear assets and sites.[17] The NCA consists of the Development Control Committee (DCC) and Employment Control Committee (ECC) (Fig 6.1). The two committees function independently for operational and developmental aspects. The Director General of the SPD is a member and Secretary of both committees.[18]

**Fig 6.1: Organisation of the National Command Authority**

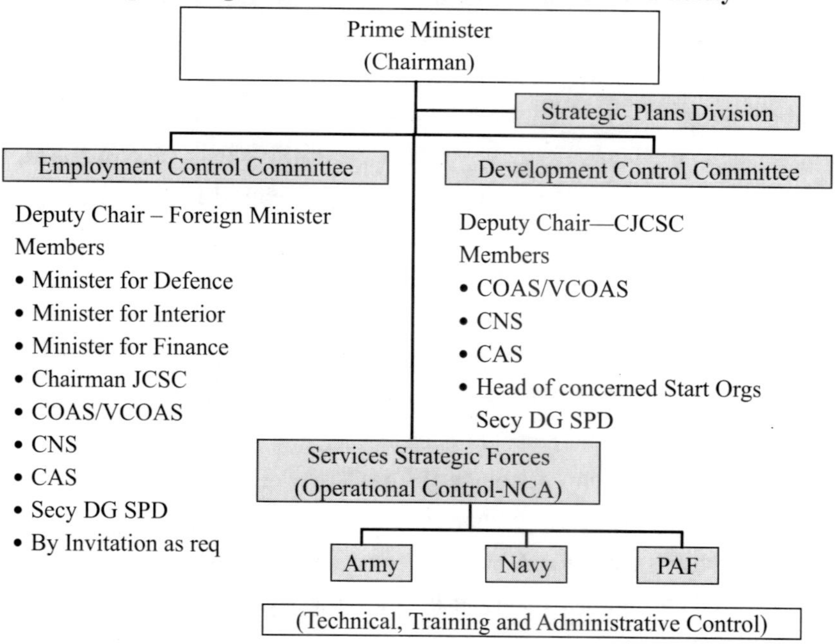


*Source:* Christopher Clary, "Thinking about Pakistan's Security in Peacetime, Crisis and War," IDSA Occasional Paper, No. 12, New Delhi, IDSA, 2010, p. 12.

***Employment Control Committee***

The ECC is the main policy-making organ and functions as a politico-military committee.[19] The ECC is the superior committee and it periodically reviews the latest information on the threats to national security, the strategic weapons programmes, their developments and deployments. It takes the decision about the responses to the developments. It provides the policy direction during peace-time and also controls the employment of the strategic forces command during the war.[20] It plays a significant role in the evolution of the nuclear doctrine and, thus, also supervises the policy for the use of nuclear weapons. It overlooks the command and control system in order to safeguard or prevent any accidental or unauthorised use of the nuclear arsenal.[21] The composition of the ECC involves representatives from both the political and military sides, providing it a good balance in some form (Fig 6.1). It includes the Foreign Minister (as the Deputy Chairman), federal Ministers for Defence, Finance and Interior, the Chairman, Joint Chief of Staff Committee (CJCSC), Chiefs of all the three Services and Director General (DG) SPD. Other officials can be a part of the committee by invitation. The ECC is responsible for employment of the Pakistani deterrent capability; it is responsible for identifying the national security threats which would lead to the use of nuclear weapons. The SPD assists the committee in its preparatory work and also makes recommendations for the approval of the committee and the NCA.[22] This arrangement potentially functions towards avoiding any accidental launch or launch through miscalculation.

The SPD head, Gen Khalid Kidwai outlined the contingencies or scenarios in which Pakistan would resort to nuclear weapons in an interview given to Italian journalists which was published in 2002. He went on to state that Pakistan would definitely use nuclear weapons in case deterrence fails and:

- If India attacks Pakistan and conquers a large part of its territory (space threshold);
- If India destroys a large part either of Pakistan's land or air forces (military threshold);
- If India proceeds to the economic strangling of Pakistan (economic strangling);
- If India pushes Pakistan into political destabilisation or creates large-scale internal subversion in Pakistan (domestic destabilisation).[23]

It can be presumed that these contingencies were talked about only after the approval of the ECC. Also, we have had statements by the Pakistani military leaders about using nuclear weapons as a "last resort".[24] But the last resort option has not been quantified and there is nothing available (at least in the public domain) which gives any specific details of the Pakistan's last resort.

***Development Control Committee***

The DCC is a subordinate committee which supervises and ensures the implementation of the decisions taken by the ECC. It is a military-technical committee (Fig 6.1) and consists of the top military leadership: CJCSC, the three Services Chiefs, and heads of various strategic organisations. The DG SPD (who is from the army) acts as the Secretary.[25]

The DCC is responsible for the development of the nuclear capabilities and keeping the deterrent in a state of preparedness. The committee exercises complete control over day-to-day technical, financial and administrative matters of the strategic organisations involved in the nuclear weapon programme, which include the PAEC, KRL, NESCOM and SUPARCO. The DCC is responsible for ensuring that the development of the strategic weapons programme progresses as per the approved development strategy.[26]

**Fig 6.2: Pakistan's Strategic Plans Division**

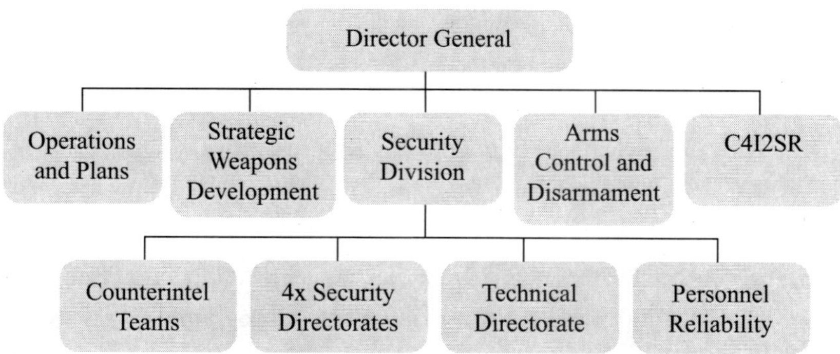


*Source:* Christopher Clary, p. 12.

**Strategic Plans Division**

The SPD is the second and most critical tier of command and control and it would not be totally incorrect to state that the entire command and control rests with the decisions, strategy and policies formulated by the SPD which is headed by the DG Army. The SPD acts as the Secretariat of the NCA, and formulates Islamabad's nuclear policy, doctrine and strategy. The SPD, rightly termed as the true "nuclear enclave," is the key to Pakistan's nuclear management.[27] The SPD reports directly to the Prime Minister. It is headed by a Director General who is appointed from the army and comprises 50-70 officers from all three Services, reflecting the army's dominant position in strategic issues.[28] The DG is entrusted with the responsibility to handle any and everything related to Pakistan's nuclear capability. The SPD is the most crucial component of the command and control structure and, in a way, manages Pakistan's nuclear capability in totality, controlling the policy formulation, operations, plans, weapons development, storage, finances and safety and security.[29] It ensures the safety and security of the nuclear assets in both the short and long terms, and helps the NCA to exercise control over the strategic organisation. According to Khalid Banuri, Director, Arms Control and Disarmament Affairs at the Strategic Plans Division in Pakistan,

> On behalf of the NCA, SPD exercises firm control over all strategic organisations and coordinates their financial, technical, developmental and administrative aspects. SPD is also responsible, in collaboration with the Foreign Ministry, for providing military inputs in the formulation of Pakistan's position on disarmament, non-proliferation and arms control issues.[30]

It is observed that the SPD under Gen Khalid Kidwai had taken firm control over Pakistani nuclear organisation and policy. The International Institute of Strategic Studies (IISS) reported, "With the SPD, the military for the first time had an organisation that could develop organisational competency and authority over the sizeable nuclear programme." The SPD carries out day-to-day management of the Pakistani strategic forces and coordinates the working of different strategic organisations involved in the nuclear weapon programme.

The SPD has four primary directorates:

- Operations and Planning Directorate, as the name suggests, carries out the operational planning.
- C4I2SR (Command, Control, Communications, Computers, Intelligence, Information, Surveillance and Reconnaissance) Directorate

is responsible for developing and maintaining strategic command and communication links.

- The Strategic Weapons Development Directorate liaises with the strategic organisations, scrutinises their budgetary demands, and carries out audits of funds.
- The Arms Control and Disarmament Affairs Directorate provides recommendations on all arms control and disarmament issues and participates in relevant bilateral and multilateral non-proliferation discussions.[31]

There are some subsidiary organisations like the Consultancy Directorate, comprising technical experts who provide technical advice on all construction projects, and the Strategic Forces Communications Planning (SFCP) cell, comprising communications experts, to assist the C4I2SR Directorate.[32]

The Security Division is the largest component of the SPD and is composed of around 9,000-10,000 personnel reporting to a serving two-star General. This division provides the internal and external security for nuclear-related sites.[33] The other directorates are: Counter-Intelligence Directorate, Technical Directorate, Personnel Reliability Programme Directorate and Security Directorate. The SPD has divisions to carry on various other functions, namely, Security Division, Civil Works Organisation and Strategic Forces Communications Planning Cell.

During peace-time, the SPD has the responsibility of guarding the nuclear programmes from external as well as internal threats. According to Luongo and Brig Salik, at present, there is a multi-layered approach to perimeter security and the SPD plays a crucial role:[34]

- **Inner Perimeter:** The central responsibility of the security of all the nuclear installations, civilian or military, is with the SPD, and the security division is the coordinating body. Reportedly, some facilities get additional protection by air defence elements.
- **Outer Perimeter:** This includes security through fencing, electronic sensors and closed circuit television cameras, etc.
- **Third Tier:** This involves counter-intelligence teams which work towards identifying the external threats.

Protection of nuclear material against theft and preventing infiltration of non-state or, for that matter, illegitimate factions, into the strategic organisations is primarily the responsibility of the SPD. The organisation does go through several programmes related to secrecy, physical security, personal screening programmes, procedural and technical controls in order to strengthen its safeguarding capabilities.[35] It is the SPD which controls the information regarding the location of Pakistan's nuclear warheads and delivery systems and, apparently, the information is not shared with the regular military and intelligence officers.

Peter R. Lavoy, who interacted with the SDP on a regular basis, has described the set-up:

> This organisation essentially coordinates with all intelligence agencies about any external threats. The Inter-Services Intelligence Directorate (ISID) forms the outermost ring of security and works closely with the security division. Prior to this, there was no formal role for the ISID in nuclear matters. Even now, the ISID director-general is not a formal member of the NCA. (Reportedly, he is a regularly invited member).[36]

Secrecy is the key to the security of Pakistan's nuclear arsenal and is certainly treated as the most significant safeguard against both internal and external threats.[37] The SPD is also responsible for the Personnel Reliability Programme (PRP). Pakistan has apparently evolved Personal and Human Reliability Programmes to ensure the entry of only authorised and trusted personnel into the nuclear installations. This factor is critical, particularly in the case of Pakistan, given the rise of extremism in the country and the attack on the Mehran Naval Base in May 2011, which clearly involved insider support. The PRP is administered by the SPD in conjunction with Pakistan's three intelligence agencies—Inter-Services Intelligence (ISI), Military Intelligence, and Intelligence Bureau.[38]

The PRP covers all persons working in the sensitive areas of the nuclear system. The PRP has been strengthened, as per Pakistani officials, much more after 9/11 and the system is based on the lessons learned and adopted from the US system. The screening involves checking of the complete background, including checks on family, political affiliations and personal inclinations. After the initial screening, there are periodic checks and screenings every two years and also at the time of transfer of an employee from one section/location to another. There is also a procedure of random checks which are carried out whenever required.[39]

The SPD coordinates with the strategic forces command. All the three forces, the army, navy and air force have their respective strategic forces command but the operational planning and control rests with the NCA.[40] The SPD formulates the short-term and long-term force development strategy for the strategic forces. The SPD is also responsible for formulating strategic and operational plans at the joint Services level for the movement, deployment, and employment of strategic forces.[41]

**Strategic Forces Command (SFC)**

The SFC is at the third tier of the command and control. All the three Services have their respective strategic forces command. The three Services retain training, technical and administrative control over their respective forces, but operational control remains under the jurisdiction of the NCA, which provides military direction through the Chairman of the Joint Chiefs of Staff Committee housed in the National Command Centre.[42] The role of the strategic forces command has been defined by Luongo and Salik:

> The primary responsibility of these commands is to exercise technical, training, and administrative control over the strategic delivery systems. The operational control, however, rests with the NCA. The army strategic force command possesses ballistic and cruise missiles, while the air force strategic command has the aircraft capable of delivering nuclear weapons. The naval strategic force command was the last to be established, and there is no public information as to whether they already have nuclear delivery systems and weapons or whether this capability is still evolving.[43]

**Decision-Making in the Command and Control Structure**

The NCA and SPD formulate the basic guidelines for the deployment and final use of the nuclear weapons. Obviously, the details of these are highly confidential and one can only draw some broad conclusions in this regard based on the various statements made by responsible military and political leaders. The decision to launch a nuclear strike rests on the consensus within the NCA, and the Prime Minister, the Chairman of the NCA would cast the final vote. Whether, the political leadership is 'actually' in a position to cast the final vote without the approval of the army for the use of nuclear weapons, is debatable. But, as per the approved command and control structure, the final authority lies with the Prime Minister of Pakistan. The NCA would then communicate the decision and delegate authority to implement the decision to the SPD and down the institutional hierarchy/structure.[44] Reportedly, Pakistan applies the two or three-man rule to the authorisation of assembly and use of nuclear weapons.[45] While, there is little clarity on the number of people involved at various steps for the deployment of the nuclear weapon, according to Pakistani officials, no single individual at any rung of the nuclear hierarchy is in a position to carry out the strike on his own and at least two or three people authenticate the launch codes for nuclear weapons.

## Measures for Safety and Security of the Nuclear Weapons Assets and Sites

### *Security of the Nuclear Weapons*

Pakistan has expanded the size of its nuclear arsenal—both the number of warheads as well as the delivery systems. Pakistan maintains three types of delivery systems—land-based missiles, cruise missiles and aircraft. The weapons are believed to be kept separate from their delivery systems, with the nuclear cores removed from their detonators. It is difficult to estimate the number of actual weapon-storage sites, but nuclear weapons would certainly be dispersed at various sites. However, Khalid Kidwai has stated that Pakistan's nuclear weapons can be mated quickly.[46] Originally, Pakistan's nuclear warheads were not equipped with Permissive Action Links (PALs), but now, as declared by Samar Mubarakmand, in an interview with a private TV network in 2004, each Pakistani warhead is fitted with a "code-lock device."[47] Pakistan follows a two-man or three-man rule to authenticate the codes that call for the release of the weapons. In 2002, Khalid Kidwai referred to a "three-man rule."[48] Later, in their writings, Feroz Hasan Khan and Naeem

Salik, mentioned the "two-man rule" that in some situations becomes a "three-man rule."[49] Neither Salik nor Feroz Hasan Khan give the details of who would be the authorised personnel for the code. It is also not clear whether these two, or in some cases three, men receive a common set of instructions through the same communication channel or different instructions through varied channels.[50] In 2006, Lt Gen. Kidwai confirmed the employment of the system equivalent to the PAL, a system used to keep a weapon from detonation in 2006, and, while addressing the US Naval Postgraduate School, he stated, "Pakistani nuclear controls include some functional equivalent to the two-man rule and Permissive Action Links (PALs) that the United States and some other nuclear-weapons states rely on."[51]

Pakistan has received assistance on the nuclear weapons security from the US and the Bush Administration reportedly spent almost $100 million on a highly classified programme to help President Musharraf secure his country's nuclear weapons, according to senior Administration officials.[52] But the US assistance does not mention the transfer of PALs, not only because the US was sensitive about sharing the information due to legal restrictions but also because the Pakistanis were suspicious that the American technology could work towards disabling their nuclear weapons.[53]

### *Safety of the Fissile Material and Spent Fuel*

The safety of fissile material is extremely critical, particularly to avoid the misuse of fissile material by the non-state actors or even by rogue members within the system. The SPD has been responsible for conducting regular audits on all nuclear inventories since 1998. Any new radioactive or nuclear material which enters the safeguard system comes under the supervision of the International Atomic Energy Agency (IAEA), which monitors the movement of the material till it is disposed.[54]

Four of Pakistan's nuclear facilities, namely, the Karachi Nuclear Power Plant and Chashma-1 power reactors and the Pakistan Atomic Research Reactors I and II in Rawalpindi, operate under IAEA safeguards. But the other important Pakistani nuclear facilities which are producing plutonium and weapon grade uranium are not under IAEA safeguards, including the Kahuta Research Laboratories and Khushab research reactor.[55]

A bigger concern for the Pakistani authorities is the safety of the spent nuclear fuel and high-activity radioactive sources which are more difficult to protect. Pakistan is making efforts to acquire specialised vehicles to prevent attempts to target transportation containers.[56] Not many details are available regarding this issue.

***Establishment of the Pakistan Nuclear Regulatory Authority (PNRA)***

On January 22, 2001, the Pakistan Nuclear Authority Ordinance, was promulgated and consequently the Pakistan Nuclear Regulatory Authority (PNRA) was established as an independent entity to implement nuclear and radiation safety regulations in the country. The PNRA is responsible for controlling and supervising all matters pertaining to the safety of nuclear installations.[57]

The PNRA is headed by a Chairman and is assisted by two full-time and seven part- time members. The DG SPD is also a member of the PNRA. The Authority functions as a regulatory and decision-making body.[58] The main functions of the PNRA are as follows:

- It takes licensee decisions for both nuclear and radiation facilities.
- It sets policies and directions on safety and environmental issues that concern the nuclear industry and the public.
- It maintains close liaison with all national organisations involved in any form in the execution of its responsibilities.
- It maintains a computerised inventory of radioactive sources in the country. The safety and security of radioactive sources are given the utmost importance in the PNRA activities.
- It is anticipated to play a significant role in the national arrangements in response to a nuclear accident or radiological emergency. The PNRA has established a National Radiation Emergency Coordination Centre (NRECC) to coordinate various steps during a nuclear emergency.[59]

In 2002, the PNRA announced a series of fresh measures for protecting "the plant and society from hazards that could be man-made or natural"[60], which included "stricter quality control and monitoring for infrastructure and equipment, multiple physical barriers to uncontrolled release of radioactive materials, radiation protection and acceptance criteria, and disaster mitigation equipment and arrangements."[61]

Radiological sources do become an area of concern and these can be misused by the rogue elements much more easily as compared to other nuclear equipment. According to Luongo and Salik, it is difficult to ascertain the total number of radiological sources in Pakistan. However, it is estimated that 65 percent of the sources are claimed to be stored and 34 percent of sources are in use. More than 50 percent of the amount in use is not under the PAEC.[62]

***Export Controls***

The internal export control guidelines for all the nuclear organisations were issued in 2000 by the SPD. Before this, the organisations were acting independently, thus, leaving numerous opportunities for suspicious transactions. The institutions are now required to follow certain procedures for all exports and seek clearance from the SPD and Ministry of Foreign Affairs.[63]

The Export Control Act was promulgated in 2004 in order to exercise export control on goods, technologies, material and equipment related to nuclear and biological weapons and their delivery systems. The salient elements of the Export Control Act include:

- Control over export, reexport, transhipment and transit of goods, technologies, material and equipment covered. Prohibition of diversion of controlled goods and technologies.
- Wide jurisdiction (also includes Pakistanis visiting or working abroad).
- Envisages an authority to administer rules and regulations framed under this legislation. Also provides for the establishment of an Oversight Board to monitor the implementation of this legislation.
- Comprehensive control lists and catch-all provisions.
- Licensing and record keeping provisions.
- Penal provisions: up to 14 years imprisonment and Rs. 5 million fine or both, and on conviction, the offender's property and assets, wherever they may be, shall be forfeited to the federal government. Right of appeal is provided for.[64]

Gen Musharraf did try and take certain steps to ensure accurate monitoring of the import of nuclear and biological weapons. But given the more than 2,000 km of Pakistan's permeable border, it is indeed challenging to monitor and control radiological import/export.

**US Assistance in Securing Nuclear Weapons**

Preventing Pakistan's nuclear weapons from a potentially disastrous possibility of falling into the extremists' hands, or an Islamist coup is certainly on Washington's priority list. The American concerns were evident in the speech of Chairman of the Joint Chiefs of Staff Admiral Michael Mullen on September 22, 2008:

> To the best of my ability to understand it—and this is with some ability—the weapons there are secure. And that even in the change of government, the controls of those weapons haven't changed. That said, they are their weapons. They're not my weapons. And there are limits to what I know. Certainly, in a worst case scenario, with respect to Pakistan, I worry a great deal about those weapons falling into the hands of terrorists and either being proliferated or potentially used. And so, control of those, stability, stable control of those weapons, is a key concern.[65]

Although Pakistan has consistently claimed the infallible security of its arsenal, it is also true that alarming incidents indicating threats to the nuclear weapons have taken place in the past. On American requests, the Pakistani authorities detained Bashiruddin Mahmood and Abdul Majeed, retired Pakistan Atomic Energy Commission (PAEC) officials on October 23, 2001. The British trained scientists had a long career in the Pakistani nuclear programme and had held multiple important positions before they retired in 1999.[66] The officials have been involved in relief work in Afghanistan through a non-governmental organisation, Ummah Tameer-e-Nau (UTN), which they had set up, and had worked on various developmental projects in Kabul and Kandahar. In November 2001, the Coalition forces in Afghanistan discovered documents that indicated the UTN's interest in biological weapons. Under interrogation, the two scientists revealed to the

Pakistani authorities that they had a series of meetings with Osama bin Laden and the last meeting took place in August 2001. They reportedly discussed with Al Qaeda about the development of chemical, biological and nuclear weapons.[67] The two men were released and arrested again, and were finally freed in 2001, without any charge. This was despite the fact that they had breached the secrecy oath in their visits to Afghanistan.[68] The United States also wanted the detention of two other nuclear experts, Suleiman Asad and Mohammad Ali Mukhtar, who are believed to have been associated with Pakistan's secret nuclear installations. The Bush Administration was suspicious of their possible links with the militants, but the two experts were never detained.[69]

The US obviously has been concerned, and has offered assistance for the safety of Pakistan's nuclear arsenal. Reportedly, the US offered nuclear security assistance to Pakistan post September 2001 which included the sharing of best practices and technical measures to prevent unauthorised or accidental use of nuclear weapons, as well as contributed to the physical security of the storage facility and personnel reliability.[70] The issue of offering PAL technology was debated in the US Administration. While many nuclear experts in the federal government favoured the offer given the fact that Pakistan's nuclear arsenal was the most vulnerable arsenal in the world, others felt that sharing of technology would teach Pakistan too much about American weaponry. Apparently, the same logic restricted the Clinton Administration from sharing the technology with China in the early 1990s.[71]

Feroz Hasan Khan in his article in *Arms Control Association* stated:

> In 2001, US Secretary of Sate Colin Powell offered nuclear security assistance to Pakistani President Gen Pervez Musharraf. The SPD carefully examined the offer and accepted training but declined technology transfers, which they perceived as intrusive or likely to compromise programme secrecy. Since then, Pakistan has benefitted from advanced-level training from US national laboratories and has improved its best practices in accordance with its own security culture.[72]

The US assistance was confirmed during Secretary of State Condoleezza Rice's confirmation hearing in January 2005. She responded to Senator John Kerry's query regarding the safety of the Pakistani nuclear weapons in the event of a radical Islamic coup in Islamabad, and stated, "We have noted this problem, and we are prepared to try to deal with it."[73] However, there is great suspicion about the US' intentions in Pakistan which aggravates each time there is a press report or comment about the US authorities taking control of Pakistan's nuclear arsenal in the wake of a breakdown of the governance which remains in constant turmoil or the prospective rise of *jihadi* elements in the leadership positions.[74] Pakistan has maintained its stance on denying any foreign access to its arsenal precisely because it fears that the knowledge might be used to target the weapons. They have also been hesitant in receiving technical assistance fearing that the US will install tracking devices to locate the weapons or secret methods of disabling them.[75]

Pakistan has been in a state of denial with respect to receiving any assistance from the US in this regard. As recently as November 8, 2009, a Pakistan Foreign Office spokesperson stated that Pakistan "does not require any foreign assistance in this regard". Pakistan will never "allow any country to have direct or indirect access to its nuclear and strategic facilities," and further added that "no talks have ever taken place on the issue of security of Pakistan's nuclear arsenal with the US officials."[76] Some press reports have indicated that the US provided Pakistan with PAL in 2003, but retired Brig Gen Naeem Salik stated, "Pakistan has developed its own PAL system which obviously ensures that even if an unauthorised person gets hold of a weapon, he cannot activate it unless he also has access to the electronic codes."[77]

The US has provided assistance in securing Pakistan's nuclear weapons, although the efforts have been restricted due to Pakistan's reluctance to provide access to the nuclear facilities. Some American experts say that "they have less ability to look into the nuclear laboratories where highly enriched uranium is produced".[78] A former US official's comment regarding the US assistance was, "Everything has taken far longer than it should..... and you are never sure what you really accomplished."[79] It is unclear as to what extent Pakistan has shared information regarding its arsenal. Also, the

Pakistani officials have stated that they have been discreet about accepting assistance and that Pakistan reserves the right to "pick and choose" the nuclear security measures, and will accept only such measures as are "non-intrusive."[80] Reports suggest that the US had spent approximately $100 million on the classified nuclear security programme by 2007, which is not so significant considering the total amount of aid Pakistan has received from the US, starting 2002, which amounts to more than $20 billion.

**Risks to the Nuclear Weapons**

Pakistan's nuclear weapons are "stored in component form", which implies that the nuclear weapons are stored separately from the delivery vehicles. Reports suggest that Islamabad's nuclear weapons are stored unassembled, with the fissile core separated from the non-nuclear explosives. Khalid Kidwai, the head of the SDP, has said that the nuclear warheads (containing the fissile cores) may be mated with their delivery vehicles.[81] Kidwai has reportedly also stated that the nuclear weapons "will be ready when required, at the shortest notice; (but) the Pakistani doctrine is not endorsing the US-USSR model with weapons at hair trigger alert". The 2001 Defence Department report suggested that Pakistan can probably assemble its weapons fairly quickly.[82]

Storage of nuclear warheads away from the delivery vehicles could prevent accidental launches, but the political and domestic situations make the nuclear weapons extremely vulnerable. When Pakistan developed its nuclear weapons, its prime concern was an attack by India, and now after 25 years of nuclearisation, Pakistan's biggest nightmare is the safety of its nuclear weapons from the *jihadi* elements on its own soil.

Pakistan has, undoubtedly, evolved a structured command and control system which is impressive. But the danger for Pakistan's weapons does not emerge from the loopholes in the command and control but from the uniquely dangerous environment in which the weapons are placed. At present, hardly a week passes by in Pakistan without a daring terrorist incident, not sparing even the armed forces, or, for that matter, the GHQ. However, it will be useful to analyse some of the scenarios in which the unauthorised use of nuclear weapons/material could occur.

***Miscalculation***

Although the final nuclear decision-making regarding the use of the nuclear weapon requires two or may be three authorised personnel to take the decision, the persistent domestic instability within Pakistan, with constant rifts between the army and the civilian forces, leading to use of nuclear weapons through miscalculation cannot be negated. A.Q. Khan acknowledged this threat:

> .... there is a real danger of nuclear war by accident due to technical failure or malfunctioning, or due to accidental detonation or launching of a nuclear weapon. Nuclear war can also be started by unauthorised action, human error or sheer madness. There is, moreover, a great danger of a person or a group of persons responsible for launching nuclear weapons going insane and deciding to launch a nuclear attack on the enemy, eliciting immediate retaliation and a real holocaust.[83]

Former Pakistani Foreign Minister Agha Shahi, had raised similar concerns:

> Pakistan may well be confronted with a hair trigger alert situation. Neither country [India and Pakistan] has an effective early warning system against missile attacks to detect intruding aircraft carrying nuclear devices. The flight time of their [ India's] nuclear-armed short and medium range missiles is only three to less than ten minutes. Hence, a 'launch-on-warning' system [launching missiles before incoming missiles arrive] that existed between the United States and the former Soviet Union—25 minutes of warning time in case of ICBMs [Inter-Continental Ballistic Missiles] and 15 minutes in that of their SLBMs [Submarine Launched Ballistic Missiles] from deployed submarines—would elude Pakistan and India. The risk of a nuclear strike by miscalculation or unauthorised use, therefore, cannot be but high, as the respective command and control systems cannot ensure real-time instructions for alerting and launching nuclear forces.[84]

The danger of miscalculation is more exasperating in the case of Pakistan where the doctrine is unwritten and consistently there has been a tendency of projection of irrationality from the governing elites. Pakistan has often threatened to go nuclear in the case of even the threat of an Indian conventional attack. In fact, the military has tried to exploit its irrational posture rather judiciously on various occasions. Practice of this nature increases the probability of accidental use of nuclear weapons.

***Outsider Threat***

The threat of the nuclear facilities being attacked by the terrorist/ extremist organisations has been most worrisome and has raised concerns both worldwide and within Pakistan. The rise of extremism in Pakistan has been significant post 9/11. Raging insurgency on the Frontier borders, Al Qaeda's leading associates hiding in the heart of Pakistan and most importantly, the rise of the Tehrik-i-Taliban Pakistan (TTP), with a strict anti-state and anti-military agenda, has made the military and nuclear installations prime targets. The nuclear installations and also the nuclear personnel have been targeted in the past on several occasions.

In 2007, two Pakistan Air Force (PAF) facilities associated with Pakistan's nuclear weapons were attacked. A suicide bomber riding an explosive-laden motocycle blew himself up near a bus carrying the air force personnel on duty to Sargodha Air Base,[85] considered to be the country's most important air base. It is the headquarters of the Central Air Command, and two squadrons of F-16 fighter jets are based there. The F-16s are supposedly Pakistan's prime aerial delivery system. On December 10, 2007, a school bus carrying the children of air force personnel outside was targeted at Kamra Air Base, which is a base for the Mirage Vs which, supposedly, could be used for nuclear missions. On August 21, 2008, twin suicide bombers carried out an attack on the gates of the Pakistan's ordnance factory at Wah, which is reportedy a site for nuclear weapons also.[86] The Pakistani Taliban claimed the responsibility for the attack.

On October 10, 2009, the TTP and Lashkar-i-Jhangvi militants attacked the Army GHQ in Rawalpindi, the most secure military complex in Pakistan. Ten TTP militants dressed in army uniforms attacked the GHQ, killing six people, including a Brigadier and a Lieutenant Colonel. According to Professor Shaun Gregory, Director of the Pakistan Security Research Unit at the University of Bradford, UK, the modalities of this attack add up to a vital blueprint for an attack on a nuclear weapons facility. For example,

- The penetration of several security checkpoints and barriers.
- Use of army uniforms, fake ID cards , a military vehicle to deceive the army.
- Use of a safe house near the GHQ for weeks before the attack.
- Use of maps of the GHQ to allow operational planning. This also projects the possibility of internal involvement in information sharing or poor management of the GHQ.
- Use of tactics that allow final tier barriers to be penetrated: grenades and/or suicide detonations at entry points which then allow penetration by follow-up commando-style groups.[87]

During the attack on the GHQ, the terrorists entered the complex and took nearly 40 people hostage. It took the Pakistan Army Commando's Special Services Group (SSG) more than 20 hours to gain control of the situation and free most of the hostages.[88]

These incidents indicate that the scenario of a militant organisation taking control of a nuclear site or some part of a nuclear site is not a remote possibility. As the number of nuclear weapons grows, the challenge to secure this enhanced number would also intensify. Militant outfits like the TTP clearly adopt an anti-state posture and have ample support from international terror outfits like Al Qaeda, with which it shares an ideological alliance. Thus, the financial and military assistance for these outfits never seems to be problem. Taliban spokesman Ehsanullah Ehsan has openly talked about using the Islamic bomb: "Pakistan is the only Muslim nuclear power state. Isn't it a shame for us to have the Islamic bomb, and even then, we are bowing down to the pressures of America?"[89]

A more likely scenario for terror outfits would be to get hold of the spent fuel which can be converted into a dirty bomb. Bharad Karnad talked about this scenario of nuclear terrorism in his article:

> The much easier route to nuclear terrorism is the 'dirty bomb'—a radiation diffusion device (RDD) that can be configured by terrorists. All that is needed is a bundle of spent fuel (featuring radio-isotopes such as cobalt 60, strontium 90 and cesium 137) powering X-ray machines, gamma ray cameras, etc., used extensively in the health and industrial sectors, a smuggled 'pencil' fuel at a time. These can be tied together and fitted into a shoebox, and the entire package blown up with a conventional explosive, such as dynamite.......a small explosion followed by release of radioactivity which, depending on ambient conditions, can spread with the wind and quickly affect an entire city.[90]

When Pakistan was developing the nuclear capability in the 1970s and 1980s, the obvious objective was to choose the location of the nuclear sites far from the India-Pakistan border in order to avoid any potential damage that can be caused by the Indian forces during war. Most of Pakistan's nuclear infrastructure is located in the north and west of the country which is either close to, even within, the region dominated by the Taliban and Al Qaeda militants (See Fig 6.3). The tribal belt of Pakistan has been home for the militants for over a decade now and this certainly adds to the concerns of safety and security of the Pakistani nukes.

### *Insider Threat*

The possibility of an insider conspiracy or insider-outsider collusion, with an objective to take over nuclear items poses a severe challenge to the command and control set-up in Pakistan. The assassination of Salman Taseer, Governor of Punjab, a critic of the blasphemy law, by his own bodyguard in broad daylight, who was then welcomed by a shower of rose petals in the judiciary, raises serious doubts about the credibility of the personnel guarding the military and nuclear establishments. Although the SPD is extremely confident about the PRP and, according to them, the programmes have a screening process from "cradle to grave," the possibility of extremist motivation leading to a mishap cannot be ruled out.

**Fig 6.3: Pakistani Nuclear Facilities**

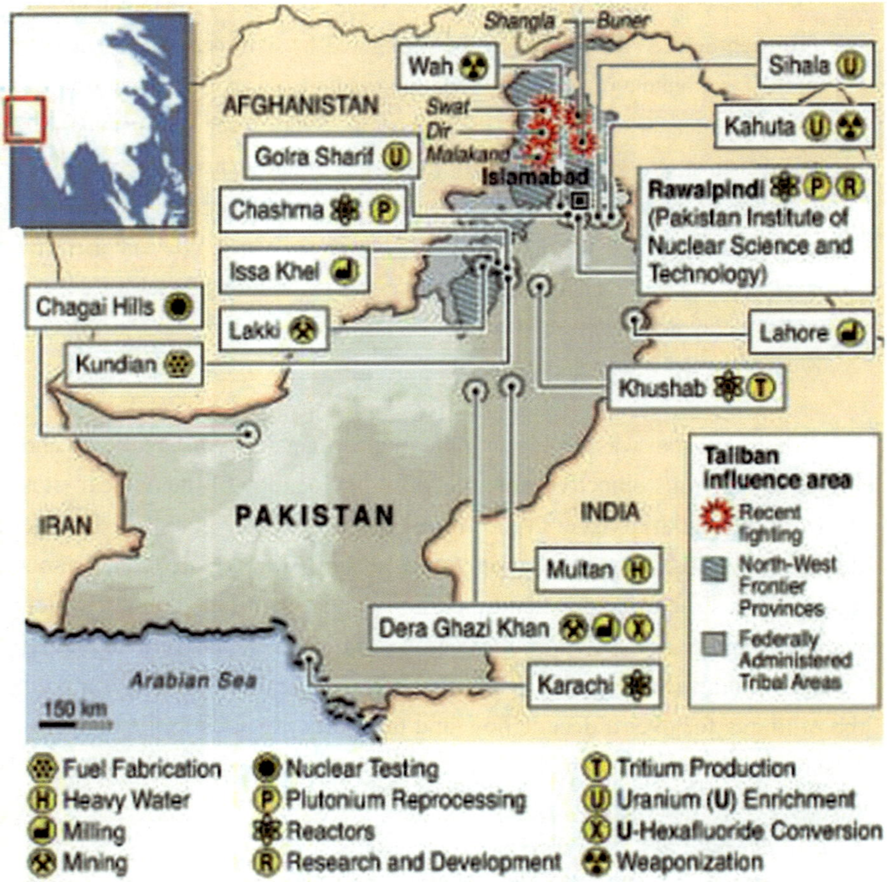


Boundary representation is not necessarily authoritative.

*Source:* Based on *Reuters*, http://blogs.reuters.com/pakistan/files/2009/08/paknuke.jpg

In May 2011, the Pakistani Taliban attacked the Mehran Naval Base in Karachi, about 15 miles away from a suspected nuclear storage facility in another glaring example of insider collaboration.[91] The Mehran Base attack could not have been possible without insider assistance. As many as 70,000 people in Pakistan have access to knowledge of some element of a nuclear facility or weapon system, given the size of the arsenal and personnel involved in the entire programme. In an environment where the survival of liberal thinking is difficult, the possibility of extremism influencing the personnel is certainly not ruled out. Dr Gregory, in his report, stated:

> As the number of nuclear weapons facilities grows, and the number of those with access to nuclear weapons or related components rises, the complex challenge of assuring the security of nuclear weapons and nuclear weapons components will become even more difficult.[92]

The insider threat may be reflected in various forms like leakage of sensitive information, assisting an outsider to break in, or theft of nuclear material. Feroz Hasan Khan, in his article, talks about misperceptions regarding the safety of nuclear weapons, but he acknowledges the insider threat:

> Western fears about Pakistani nuclear security range from valid to bizarre. The more valid concerns involve theft of material, sabotage, unauthorised use of nuclear weapons, and insider-outside collaboration.... The most difficult aspect of measuring effectiveness comes from the unpredictability of human motivations. Motivated individuals can always elude barriers.[93]

Sig Hecker, in his article, states:

> Pakistan is also at the top of my list of nuclear risks because it is the most likely place in which fissile materials could find their way out of the hands of government and into those of terrorists. However, it is also the most difficult place to do science diplomacy. The five-nation dialogue facilitated discussion with Pakistani officials on nuclear materials security and the proliferation activities of Pakistani nuclear scientist Abdul Qadeer Khan. But Pakistan's nuclear scientific community is not accessible, and the problems are difficult and extremely sensitive politically....[94]

With the expansion of the arsenal, the nuclear expertise is also expanding within Pakistan and the challenge for the military and the civilian regime is of controlling the nuclear expertise. Although, Pakistan has reemployed the nuclear scientists in various other programmes, dealing with the retired scientists is certainly a problem. Protection of sensitive information is one of the biggest issues in Pakistan where the temptation could not be only ideological but also monetary.

## Conclusion

Given the volatile political situation and existing threat of *jihadi* elements within Pakistan, the issue of safety and security of the nuclear weapons is certainly one of the biggest concerns for not only the Pakistani establishment but also the international community, particularly the United States. Post 9/11, Pakistan has worked towards strengthening its nuclear command and control system and has received over $100 million of US assistance for increasing the security of the nuclear weapons. There have been major changes in the NCA, integration of command and control, employment of the permissive action links on nuclear weapons and export control measures. The political crisis in 2007 followed by steadily increasing terrorism and raging insurgency did perk up the issue of safety of the nuclear weapons. Although the Pakistani leadership has repeatedly expressed confidence about the security of the weapons, but with the political instability, deteriorating civil-military relations and declining US-Pakistan relations, the concerns for the safety and security of the world's fastest-growing nuclear stockpile would continue to grow.

## Notes

1. See Hans M. Kristensen and Robert S. Norris, "Pakistan's Nuclear Forces, 2011," *Bulletin of Atomic Scientists*, Vol. 67 No. 4, July/August 2011, pp. 91-99.
2. Maj Gen Muhammad Ali Durrani, "Pakistan's Strategic Thinking and the Role of Nuclear Weapons", Cooperative Monitoring Centre Occasional Paper 37, Sandia National Laboratories, July 2004, p. 27.
3. "Nuclear Weapons Research, Development, Testing, Production and Naval Nuclear Propulsion Facilities: Pakistan," NUCLEAR Files.Org, at http://www.

nuclearfiles.org/menu/key-issues/nuclear-weapons/issues/facilities/pakistani-nuclear-weapons-facilities.htm.

4. Mushahid Hussain, "Media Off Traget with Pakistan Nuclear Scare", *Asia Times*, November 7, 2001, as cited in "Nuclear Black Markets: Pakistan, A. Q. Khan and the Rise of Proliferation Networks: A Net Assessment," *IISS Strategic Dossier* (UK, IISS, 2007).
5. Gordon Corera, *Shopping for Bombs: Nuclear Proliferation, Global Insecurity and the Rise and Fall of the A.Q. Khan Network* (London: C.Hurst & Co (Publishers) Ltd, 2006), p. 50.
6. Ibid.
7. Bhutto visited Washington in 1989 and according to the intelligence experts, she had no real influence over the programme. The CIA officials acknowledged in a secret paper, written just before the visit, that Bhutto was "unlikely to gain control over nuclear decision-making anytime soon." CIA Internal Report, "CIA Near East and South Asia Review: Prime Minister Bhutto and the Pakistan Nuclear Issue," May 5, 1989, declassified September 1, 2000, at www.cia.gov, as cited in Corera, n. 5, pp. 50-51.
8. Owen Bennett Jones, *Pakistan: Eye of the Storm* (New Delhi: Viking, 2002), p. 210.
9. Ibid.
10. Naeem Salik, *The Genesis of South Asian Nuclear Deterrence: Pakistan's Perspective* (New York: Oxford University Press, 2010), p. 234.
11. Ibid.
12. Christopher Clary, "Thinking about Pakistan's Security in Peacetime, Crisis and War", IDSA Occasional Paper, No. 12, New Delhi, IDSA, 2010, p. 11.
13. Paul Kerr and Mary Beth Nikitin *CRS Report for Congress*, "Pakistan's Nuclear Weapons: Proliferation and Security Issues," February 23, 2010, p. 8. Updated June 20, 2008.
14. December 2007 Ordinance to provide for the constitution and establishment of the National Command Authority. As cited in Kerr and Nikitin, Ibid., p. 9.
15. Ibid.
16. Ibid.
17. Kenneth Luongo and Naeem Salik, "Building Confidence in Pakistan's Nuclear Security", *Arms Control Today,* December 1, 2007.
18. Rabia Akhtar and Nazir Hussain, "Safety and Security of Pakistan's Nuclear Assets," in Usama Butt and N.Elahi, eds., *Pakistan's Quagmire: Security, Strategy and the Future of the Islamic-Nuclear Nation* (New York: Continuum, 2010), p. 179.
19. Luongo and Salik, n. 17, pp. 12-13.

20. Zafar Iqbal Cheema, *Indian Nuclear Deterrence* (Oxford Publishers, 2011), p. 181.
21. Ian Bremmer and Maria Kuusisto, "Pakistan's Nuclear Command and Control: Perception Matters," *SASSI Research Report,* August 2008, p. 10.
22. Cheema, n. 20, pp.181-182.
23. As cited in Paolo Cotta-Ramusino and Maurizio Martellini, *Nuclear Safety, Nuclear Stability and Nuclear Strategy in Pakistan,* Landau Network – Centro Volta Report, Rome, Italy, January 21, 2001.
24. President Pervez Musharraf, in May 2002, said that if it came to war between the nuclear-armed rivals, he would "respond with full might." Also see, Agha Shahi, Zulfiqar Khan and Abdul Sattar, "Securing Nuclear Peace", *The News* (Islamabad), October 5, 1999.
25. Khalid Banuri and Adil Sultan, "Managing and Securing the Bomb", *Daily Times,* May 30, 2008.
26. Air Commodore Khalid Banuri's lecture on Pakistan's command and control structure at Fatima Jinnah Women University Rawalpindi, in April 2009, as cited in Akhtar and Hussain, n. 18, p. 180.
27. *IISS Strategic Dossier*, n. 4.
28. Ibid., p. 112.
29. Akhtar, n. 18, p. 181.
30. Ibid.
31. Luongo and Salik, n. 17, p. 13.
32. Ibid.
33. Clary, n. 12, p. 12.
34. Luongo and Salik, n. 17, p. 14.
35. Clary, n. 12, p. 13.
36. Peter R. Lavoy, "Islamabad's Nuclear Posture: Its Premises and Implementation", at kms1.isn.ethz.ch/serviceengine/Files/ISN/47509/.../en/Chapter+5.pdf.
37. Clary, n. 12, p. 13
38. Ibid.
39. Luongo and Salik, n. 17, p. 14.
40. Ibid., p. 9.
41. Cheema, n. 20, p. 186.
42. *IISS Strategic Dossier*, n. 4, p. 112.
43. Luongo and Salik, n. 17, p. 13.
44. Bremmer and Kuusisto, n. 21, p. 11.
45. Mahmud Ali Durrani, as cited in Ibid., p. 11.
46. Kerr and Nikitin, n. 13, p. 13.
47. Cited in. Luongo and Salik, n. 17, p. 13.

48. Clary, n. 12, p. 15.
49. Ibid., p. 15.
50. Ibid., p. 15.
51. Lt Gen Khalid Kidwai's CCC Address, "Pakistan's Evolution as a Nuclear Weapon State", Centre on Contemporary Conflict, Naval Postgraduate School, at http://www.nps.edu/academics/centers/ccc/news/kidwaiNov06.html
52. David E. Sanger and William J. Board, "US Secretly Aids Pakistan in Guarding Nuclear Arms," *The New York Times,* November 18, 2007.
53. Ibid.
54. Luongo and Salik, n. 17, p. 13.
55. Ibid. In addition there are two facilities which are under IAEA safeguards: Hawks bay Depot, Karachi and Chashma Nuclear Power Plant – 2 (under construction), in Salim Khan, M. Sayeed Mullah and Sohail Qayyum, "IAEA Safeguards in Pakistan and Emerging Issues/Challenges", IAEA-CN-184/77.
56. Ibid.
57. *Country Report on Control of Radiation Protection Practices in Pakistan, Pakistan Nuclear Regulatory Authority (PNRA)* IAEA/RCA MID TERM REVIEW MEETING IAEA MODEL PROJECT RAC/9/026, RAS/9/027 Beijing, June 7-11, 2004 , at http://www.rca.iaea.org/members/Projects/RAS9029/Appendix%20%2018A%20Country%20Report%20Pakistan.doc.
58. Ibid.
59. Ibid.
60. Luongo and Salik, n. 17, p. 15.
61. Ibid.
62. Of the total in use, 49 percent is under the PAEC, of which 26 percent is for medical use and 74 percent for non-medical use, and 51 percent is non-PAEC, of which 12 percent is for medical use and 88 percent is for non-medical use. Cited in Luongo and Salik, n. 17, p. 16.
63. Luongo and Salik, n. 17, p. 15.
64. Pakistan's national legislation entitled: "Export Control of Goods, Technologies, Material and Equipment Related to Nuclear and Biological Weapons and their Delivery Systems Act, 2004" *IAEA*, Information Circular, INFCIRC/636, November 23, 2004.
65. Kerr and Nikitin, n. 13, p. 1.
66. Zahid Hussain, *Frontline Pakistan: The Path to Catastrophe and the Killing of Benazir Bhutto* (London: Penguin Books, 2008), p. 154.
67. "Nuclear Experts Briefed bin Laden, Pakistan says," *Washington Post,* December 12, 2001, as cited in Hussain, n. 66, p. 155.
68. Ibid.
69. Ibid., p. 156.

70. Joby Warrick, "US Has Concerns Over Security of Pakistan's Nuclear Weapons," *The Washington Post*, November 11, 2007; David Sanger and William Broad, "US Secretly Aids Pakistan in Guarding Nuclear Arms", *The New York Times,* November 17, 2007, as cited in Kerr and Nikitin, n. 13, p. 13.
71. Sanger and Broad, in Ibid.
72. Feroz Hasan Khan, "Nuclear Security in Pakistan: Separating Myth From Reality", *Arms Control Association,* July/August 2009.
73. "The Nomination of Condoleeza Rice to the Secretary of State," Hearings before the Senate Foreign Relations Committee, January 18 and 19, 2005, as cited in Karr and Nikitin, n. 13, p. 13.
74. Michael Krepon "How Safe and Secure are Pakistan's Nuclear Weapons?" June 13, 2008, Condensed version of the testimony submitted to the Senate Committee on Homeland Security and Government Affairs on June 12, 2008, at http://beta.stimson.org/pub.cfm?ID=628
75. Gordon Corera "How Secure is Pakistan's Bomb?" February 4, 2008, BBC News, at http://news.bbc.co.uk/2/hi/south_asia/7225175.stm
76. "Pakistan Foreign Office Rejects US Media Report on Nuclear Arsenal", *Associated Press of Pakistan*, November 9, 2009, as cited in Kerr and Nikitin, n. 13, p. 13.
77. Corera, n. 75.
78. Sanger and Broad, n. 71.
79. Ibid.
80. Kerr and Nikitin, n. 13.
81. Maurizio Martellini, "Security and Safety Issues About the Nuclear Complex: Pakistan's Standpoints. A Concise Report of a Visit to Islamabad by Landau Network Centro Volta (LNCV) Mission, carried out on February 9-13 2008" as cited in Kerr and Nikitin, n. 13, p. 10.
82. "Proliferation :Threat and Response : January 2001," US Department of Defence, p. 27.
83. Quoted in Hussain and Kamran, eds., *Dr A.Q Khan,* p.28, as cited in Jones, n. 8, p. 207.
84. Shahi, address to seminar, Islamabad, February 2000, as cited in Jones, n. 8, p. 208.
85. "Nine Killed in Suicide Attack on PAF Bus", *Pak Tribune,* November 1, 2007, at http://paktribune.com/news/Nine-killed-in-suicide-attack-on-PAF-bus-193375.html
86. See Sabir Shah, "40 Major Terror Hits Rock Armed Forces Security Agencies," *The News*, May 24, 2011.
87. "Terrorist Tactics in Pakistan Threaten Nuclear Weapons Safety", *Combating Terrorism Centre,* June 1, 2011.

88. Ibid.
89. Bharat Karnad, "5 Minutes to Nuclear Jihad?," *Asian Age*, May 29, 2011.
90. Ibid.
91. "Insiders Helped Militants Plot PNS Mehran Naval Base Attack: Experts," *The Nations*, May 23, 2011.
92. Peter Goodspeed, "Pakistan's Nuclear Arsenal may be 'Compromised': Report," *National Post*, June 14, 2011, at http://news.nationalpost.com/2011/06/14/pakistans-nuclear-arsenal-may-be-compromised-report/
93. Feroz Hasan Khan, "Nuclear Security in Pakistan: Separating Myth From Reality," *Arms Control Today*, July/August 2009.
94. Siegfried S. Hecker, "Adventures in Scientific Nuclear Diplomacy", *Physics Today*, July 2011, at http://www.physicstoday.org/resource/1/phtoad/v64/i7/p31_s1?bypassSSO=1

# 7. Future of Pakistan's Nuclear Strategy

Pakistan's defence planners have essentially relied on nuclear weapons for the security and ultimate survivability of the state. The basic assumption behind nuclear weapons has been that they will not only guard Islamabad against external aggression, but, would also protect its ideological identity as the only Muslim nation possessing nuclear weapons. However, the most imperative rationale behind the build-up of the nuclear weapons has been Pakistan's desire to at best match India's military superiority, if not neutralise it strategically. Pakistan's desire for high technology weaponry in the 1950s led to its joining the Western alliance against the Communist expansion, in less than a decade after its creation in 1947. The perceived threat from India has perpetually dominated defence planning in Pakistan. On the other hand, even with the deep sense of insecurity and perceptible Indian conventional superiority, Pakistan launched three wars against India and its sense of insecurity deepened with the experience of defeat in each of the wars.

The drive for nuclear weapons was initiated by Zulfiqar Ali Bhutto after Pakistan's humiliating defeat in the 1971 War. Although Bhutto was inclined towards the nuclear weapons even in the 1960s when he served under Ayub, Pakistan was reluctant to adopt the nuclear route as long as Ayub ruled. But after December 1971, Z. A. Bhutto felt free to initiate a nuclear weapon programme through clandestine means and extensive support from China, and was even facilitated by the US in the early 1980s. Gen Zia continued the efforts for the acquisition of the bomb despite several pressures from the West. For the Pakistani leadership, nuclear weapons were the only answer to the question of maintaining national unity and to protect further disintegration of Pakistan.

The acquisition of the nuclear weapon capability also seems to be critical for Pakistan, for portrayal of its image as the leader of the Islamic world and, thus, its appeal for the 'Islamic bomb' has been strong, enabling it to gather sufficient funds from the richer Muslim countries in the 1970s and 1980s in support of the nuclear bomb programme. In its pursuit for nuclear power status, Pakistan not only acquired funds from Saudi Arabia and Libya, but also shared the nuclear expertise with Iran, Libya, and Iraq.

Pakistan has relied for more than a quarter century on its nuclear weapons to conduct its strategy against India. The military's grand strategy has rested on maintaining the centrality of the covert war (war through terrorism) strategy. Pakistan's acquisition of nuclear weapons has been rationalised as a deterrent to Indian conventional military superiority and, hence, to provide an umbrella under which to pursue a proxy war through terrorism. Nuclear weapons have been perceived as a shield to deter an Indian aggression which possibly could be prompted in response to the provacative (sustained and/or high profile) terrorism being conducted from the Pakistani soil, sponsored by the state actors. Pakistan's reliance on nuclear weapons has also increased with its growing instability and the rising threat of failure of the state.

Pakistan had acquired complete nuclear capability in 1983, and by 1987, it was believed that it could carry out a full nuclear explosion. After the overt nuclearisation in 1998, Pakistan has continued to expand its nuclear arsenal along with the delivery systems. Presently, it has the fastest-growing stockpile in the world and it is estimated that in 2011, Pakistan had a stockpile of 90-110 nuclear warheads, which is a steep increase as compared to 70-90 warheads which it possessed in 2009. [1]

Now the question one needs to address is: what would be the future of this expanded arsenal? Two aspects in this context need to be evaluated. Firstly, how safe is Pakistan's nuclear arsenal, given the current instability and predictable uncertainty within Pakistan. Secondly, what would be the future of Pakistan's nuclear strategy? In other words, how would it potentially plan to use its nuclear weapons in the foreseeable future?

The safety and security of this rapidly growing arsenal is one of the most critical issues that has been a matter of intense debate, and has raised numerous concerns worldwide. In the past, Pakistan's proliferation has shown the failure of the state's command and control over its nuclear arsenal and also its lack of responsibility in terms of handling of the nuclear technology. Especially if we go by the Pakistan government's posture that proliferation from Pakistan to countries like Iran, Libya and North Korea was undertaken single-handedly by Dr A. Q. Khan, and the international community, led by the USA, believed and acted on this assumption. Although Pakistan has managed to evolve a sound command and control system and has repeatedly expressed confidence in the security of its arsenal, given the current instability in Pakistan, qualms regarding misuse or accidental use of its arsenal are inevitable. In this context, it would be useful to have a look at the current state of Pakistan.

**State of Pakistan**

The domestic stability and societal cohesion of a state and the nature of its domestic power structure would inevitably have a dominant influence on its nuclear arsenal in all its dimensions, especially, the confidence of the international community in general and of Pakistan's neighbourhood in particular. It is obvious that the Pakistan Army has controlled the major and critical components of the nation's power structure since the mid-1950s, including the periods during which it did not exercise direct political control and rule. One of these critical areas has been the nuclear weapons capabilities. In this field, the army leadership has not even recognised the legitimate role of the other two Services, knowing fully well that the Pakistan Air Force surely (and Pakistan Navy) would have to play a natural role in the delivery of nuclear weapons if not in various aspects of decision-making related to possession and employment of nuclear weapons. The struggle by the army to retain total and proprietary control over the nuclear arsenal came to the force much more after the nuclear test in 1998. The army has managed retain the chairmanship of the Joint Chiefs of Staff Committee (JCSC) since 1998. Since the time of the creation of the JCSC in 1976, the air force and navy have chaired the committee only once and twice respectively.

But we must note that the armed forces of any country, though trained and equipped to fight, remain a reflection of the society at large, and its values. The armed forces and their components inhabit and build their own institutional culture and personality. But the broader national society's culture and personality undergo many socio-economic changes which would inevitably influence the people who constitute the armed forces. Islamisation of Pakistan is a vivid example, which involved the Islamisation of the army (initiated by Gen Zia), which embraced *jihad* as the motto of the army and also, the army endorsed the "Quranic Concept of War" [2] legitimising "terrorism" and "total war"—both central to the employment of nuclear weapons. Today, religious extremism and the *jihadi* factor are seen by the international community as the great danger, in terms of the security and safety of Pakistan's nuclear arsenal for which the US has provided monetary and technological help to reduce the potential dangers.

It is in this context that we need to focus at least briefly on Pakistan's domestic stability and societal cohesion of state:

Pakistan today is perhaps at the worst stage of its destabilisation since its inception. Alarming radicalisation, lowering GDP, rising inflation, exceeding anti-state and particularly anti-military sentiments, seem to have wracked the nation and the very survival of the state is being questioned. The US' greatest success in the war on terror, the killing of Osama bin Laden within the Pakistan military cantonment, next to a military academy, has raised a number of questions regarding not only the credibility of the Pakistani leadership but also its future relationship with Washington which would further impact the stability quotient in Pakistan. The already complex problems in Pakistan have been compounded by the lowering US popularity in Pakistan, where not only the leadership but even the common man blames the decade-long US presence for much of its woes and also is apprehensive about Western intentions to grab its nuclear weapons. The national interests of both Pakistan and the United States appear to be in complete opposition but neither of the two has overtly admitted this as they complement each other's requirements right now.

### *Political Imbroglio*

Pakistan has never been politically stable and never in the history of Pakistan has a democratically government served its full term and been replaced by another democratically elected government. The military's role in the political structure which commenced with the proclamation of the first Martial Law has continued to expand and it has emerged as the most influential political party within Pakistan.

The year 2007 witnessed the military willingly taking a back seat from Pakistan's political front, and the unprecedented rise of civilian forces, indicating a shift in the civil-military balance in the country. The judiciary came out as an extremely potent force which could create compelling circumstances for the Army Chief to reverse his decision by reinstalling Chief Justice Choudhary. The 2008 elections were obviously very controversial and their credibility was debatable. These elections generated innumerable assumptions but the two major political parties along with other small parties were, for the first time, able to put their differences aside and come together to revive democratic rule in Pakistan. It was evident that much of this rise of the political forces in 2008 originated from the strong anti-military and, more specifically, anti-Musharraf sentiments in Pakistan. The much awaited civilian elected leadership did come up in Pakistan and the military did opt to step down from direct political control, but the civilian control in Pakistan has its limitations. The PPP which came into power swinging the sympathy votes, failed to deliver on most of the promises it had made during the elections and has been widely accused of corruption. The army has little fondness for Zardari but the country's most powerful institution had little choice as the primary rival to Zardari, the alternative for the army, was Nawaz Sharif whose regime tumbled in 1999 after he sacked the former Army Chief Jehangir Karamat and attempted to sack Gen Musharraf.

Although the PPP came into power, it has been an irritant to the men in khaki and the rift between the two was not unpredictable. Zardari was locked in a bitter face-off with Kayani, bringing Pakistan to the brink of yet another military takeover in the wake of the "Memogate" issue[3] which erupted after Osama's killing in Abbottabad in May 2011. Tensions between the civilian leadership and the military reached its peak when Prime Minister Gilani accused the army chief and ISI chief of acting 'unconstitutionally' and 'unlawfully' by directly testifying before the Supreme Court in the Memogate case. The Prime Minister also sacked the Defence Secretary, a retired three-star General who is reportedly close to the Army Chief Gen Kayani. Events took their own course and the tussle of power dominance allowed the military, that had warned the civilian government of "grave consequences," to, once again, emerge as the dominant decision-making force in Pakistan. At the height of the tensions, the Prime Minster told the Parliament that he was answerable not to any individual but to the entire nation. This is not usual in the military dictated democracy and the last time such a statement was made was in 1985 by Prime Minister Junejo, during Zia's regime. And Junejo's regime met an unexpected end in 1988, when Gen Zia dismissed his government using the controversial Article 58(2) b of the Constitution.

The probability of a coup has receded for some time to come but that doesn't in any way imply that Pakistan is politically stable. The PPP would not want an early election and also the army in Pakistan realises that the time is not right for a military coup as the international environment is not conducive for a coup. The relationship between the army and government is based on mistrust which reached its peak during the Memogate scandal.

***Credibility of the Military***

Although the military remains the dominant force, the credibility of the military has suffered a major blow in Pakistan and the circumstances surrounding Osama's killing have compelled the Pakistanis to rethink their trust in the military. This raises doubts within Pakistan of the army's ability to safeguard its expanding nuclear arsenal. The fact that Osama Bin Laden was able to live for years undetected in Abbottabad has been a severe blow to Pakistan's image. The world's most wanted terrorist was comfortably placed in a complex which is home to three Pakistan Army regiments, and thousands of military personnel. Osama's killing has tarnished the military's image within Pakistan. It is difficult for the US and for the rest of the world to believe that Osama did not enjoy the patronage of the military and intelligence agencies in Pakistan in a military cantonment close to Islamabad. President Obama's chief counter-terrorism adviser, John Brennan, said it was "inconceivable that Osama bin Laden did not have a support system in the country that allowed him to remain there for an extended period of time".[4]

The esteem of the military has probably reached its lowest level in Pakistan's history. Reports suggest that the army is criticised openly in public places, without the fear of punishment. The Pakistanis are embarrassed that their supposedly all powerful military, the institution which has been a central player in the country, was unaware of the major operation on their soil. For decades, the military has highlighted the threat of Indian aggression and justified the defence build-up which has been at the cost of national development. A major chunk of national resources has been spent on defence and intelligence build-up, and every financial year, defence spending is raised substantially, citing the Indian threat. Still, the military was unable to stand up to a handful of foreign intruders and defend its own sovereignty.

The military has been the most stable institution in Pakistan which has very successfully demonstrated its capability of crisis management. In the past, Pakistanis have looked up to the military on several occasions for resolving their woes which have repeatedly surfaced due to the corrupt civilian leaders. But the frequent attacks on the military installations, including the one on the Army GHQ in 2009, Osama's killing, followed by a terrorist attack on Mehran, which reportedly had insider support, have tarnished the military's credibility. There is also dissatisfaction and frustration amongst the junior level of the military which is inevitable.

***Rising Militancy***

Presently, Pakistan is facing a broad mosaic of militancy. A vast variety of terrorist groups operate from Pakistan and share a varied relationship with the state (see Table 7.1) Some of these, the military is willing to target, some it was compelled to target, and, a few it wants to protect. For the first time, the militants have managed to penetrate the interiors of Punjab and Sindh, and the Army GHQ became one of the terror targets in the recent past. The impact of the Afghan War has allowed Al Qaeda, the Haqqani network and the Taliban inside Pakistan where they have been trying to expand their influence. The Pakistan military and the ISI have nurtured some of the terrorist organisations which they categorically grade as their "strategic assets" and are clearly not willing to act against them. These so called "friendly"[5] groups, as Ayesha Siddiqa Agha terms them, include the Good Taliban, Lashkar-e-Taiba (LeT), Jamaat-ud-Dawa (JuD), and Jaish-e-Muhammad (JeM). The Haqqani network, which has strongholds in the bordering Afghan districts of Paktia and Khost and is involved in the anti-U.S and anti-NATO operations, has enjoyed the state patronage for long.[6] The military has desisted from acting against these groups and has been "selective and partial" in its counter-terrorism approach, which has been the root cause for the lack of success in the war on terror.

**Table 7.1- Militant Groups in Pakistan**

| Confederations | Militant Groups | Targeting Scope | Pakistani Tolerance |
|---|---|---|---|
| Pakistani Taliban (Tehrik-i-Taliban Pakistan, TTP) and affiliates | Collection of semi-autonomous regional networks across FATA under emir Hakimullah Mehsud.<br>Constituent groups include Mehsud factions; TSNM; Lashkare-Islam; Wazir factions, and other local groups<br><br>Affiliates include Punjabi Taliban (splinter elements of Kashmiri jihad and sectarian groups); Ghazi Force | Pakistani state<br>Anti-Shia<br>Western interests<br>Criminal activity | None |
| Global Jihadis | Al-Qaeda; Brigade 313; Lashkare-Jhangvi (LeJ); Lashkar-e-Taiba; Tehrik-i-Taliban | Pakistani state<br>Western interests<br>Indian interests<br>Afghan interests (low capacity) | Low to medium |
| Afghan Taliban | Quetta Shura Taliban (QST); Haqqani Network (HQN); Hizb-iIslami | Western interests<br>Afghan Security Forces<br>Criminal activity | Medium to high |

| | | | |
|---|---|---|---|
| Sectarian | Sipah-e-Sahaba Pakistan (SSP); Lashkar-e-Jhangvi (LEJ); alQaeda; Islamic Movement of Uzbekistan (IMU); Jaish-eMohammad (JEM); Sipah-eMohammad (SMP) | Anti-Shia<br>Intra-Sunni<br>Anti-Sufi (limited degree)<br>Anti-Sunni (only SMP) | Medium |
| Kashmir Jihadis | Lashkar-e-Taiba (LET); Jaish-eMohammad (JEM); Harkat-alJihad-al-Islami (HUJI); Harkat-alMujahideen (HUM) | Indian interests | High |
| Baloch Separatists | Baloch Liberation Army (BLA); Balochistan Student Organization (BSO); Balochistan Republican Army (BRA); Balochistan People's Liberation Front (BPLF); Popular Front for Armed Resistance (PFAR) | Pakistani state<br>Anti-Punjabi<br>Criminal activity | None |

*Source:* Based on Anthony H. Cordesman, Arleigh A. Burke Chair in Strategy and Varun Vira, *Pakistan: Violence Vs Stability A Net Assessment*, Centre for Strategic and International Studies, Final Draft 7 June 2011, pp 43-44, at http://csis.org/files/publication/110607_Stabilizing_Pakistan.pdf

The frontiers borders – Northwest Frontier Province (NWFP) and Khyber Pakhtunwa (KPK)—have become the breeding ground for terrorism and the region presents a conglomeration of various insurgents and terrorists groups to the Pakistani military and the NATO forces. Al Qaeda, Haqqani and the Taliban have established their sanctuaries in these area. The NWFP and KPK which comprise a troublesome region even in the best of times, are now embroiled in a full blooded insurgency. Insurgency in Baluchistan, which became active in 2004, has aggravated much more after the killing of Akbar Bugti in 2007. The Baluch have long suffered a deep sense of alienation from the state and have frequently resisted it, which the national leadership tried to curb with brute military force in the past, starting in the 1940s. Added to this deadly mix of extremists is the Tehrik-i-Taliban Pakistan (TTP) whose avowed aim is to revolt against the presence of US and NATO forces within Pakistan and establishment of *Sharia* as the rule of law within Pakistan. The TTP has taken up the battle against the Pakistani state and has been very selectively targeting the army. The TTP probably tops the list of deadly terrorist attacks, including the assassination of Benazir Bhutto in 2007 and the attack on the Marriott Hotel in Islamabad in 2008. Although some Pakistani scholars hold the view that the TTP is losing its 'brand value', the fact remains that the attacks by the TTP do pose a significant challenge which the Pakistan Army is trying to counter militarily but not much success has been achieved.

The present situation in Pakistan where there is complete lack of trust between the political and military forces, the waning credibility of the country's most powerful and stable institution, and the raging insurgency, certainly pose potential threats to the vast and expanding Pakistani nuclear arsenal. The physical security of the nuclear arsenal becomes even more critical as a significant number of the nuclear sites are located in the areas (like in the Frontier borders, which was done intentionally to keep them away from India) which are home to not only the Taliban insurgents but also the global terrorist network.

Pakistan's relationship with the US has been strained and is expected to worsen in the future. There is tremendous anti-American sentiment within Pakistan and, surprisingly, for the first time, the anti-American sentiment has surpassed the anti-Indian sentiment. The Pakistanis blame the US for the spillover of violence into their country and a majority views the American aid as sheer compensation for the hardship and killing the nation has gone through due to America's war on terror. Although, the Americans have never been bothered about the civil society's sentiments, they have been comfortable dealing with the military and have pampered the military, their trust in the military has also been shaken, and the secret operation of Osama's killing was an ultimate manifestation of the mistrust. The suspension of US assistance would have security and economic implications for Pakistan. It would put on hold the US equipment and training coming into Pakistan and also the developmental programmes based on the US funding may be put on ice. Also, the US moving away from Pakistan would impact Pakistan's ability to negotiate in the international financial institutions like the International Monetary Fund (IMF) and World Bank.

**Nuclear Posture and Strategy**

The central logic around the build-up of the nuclear weapons has been to neutralise the conventional capability of the much stronger adversary, India, which according to Pakistan has hegemonic ambitions and is intolerant of a much smaller and weaker Pakistan. With this basic assumption, the declaration of the "first use" doctrine was no surprise. Pakistan's nuclear doctrine in this respect has followed NATO's nuclear doctrine during the Cold War. NATO confronted the challenge of deterring a conventionally superior capability of the Warsaw Pact and adopted a "first use" doctrine.

Pakistan till today, after fourteen years of overt nuclearisation, does not have a written doctrine because it believes that an unwritten doctrine would help in maintaining robust deterrence. However, the policy-makers have from time to time talked about the basic tenets of a doctrine which include: Indo-centricity, credible minimum deterrence and first use. Pakistan has been extremely vocal and clear about its command and control system and also to some extent about its delivery systems, but the doctrine has been intentionally kept ambiguous.

As nuclear weapon are supposed to deter India from using conventional force and to serve the objective of "war prevention", Pakistan has maintained a posture of ambiguity and kept the various options open in terms of (i) when to use nuclear weapons; and (ii) what to use, keeping its options open, from tactical nuclear weapons, the recently developed NASR to long range ballistic missiles like the Ghauri and Shaheen II, besides aircraft and possibly a warship (for sea launched missiles) . The political and military leadership has off and on declared on various occasions (especially, before the overt nuclearisation in 1998) that Pakistan would be ready to use nuclear weapons even at the slightest degree of insinuation of Indian conventional aggression and, thus, have projected a low nuclear threshold for Pakistan. The SPD head, Khalid Kidwai's interview published in 2002, created ample pandemonium as it drew vast and vague parameters for Pakistan, allowing it to use nuclear weapons first in the wake of multiple actions which India could take in the event of a crisis. Ironically, roughly during the same time and in the succeeding years, Pakistan's declaration of using nuclear weapons as a "last resort" began to surface before Pakistan started to portray a posture of a "restrained and responsible" nuclear power. This option of last resort and then restraint and responsibility projected a high nuclear threshold.

Although Pakistan talked about using nuclear weapons as the last resort, it did not hesitate to issue threats during the 1999 Kargil War or Operation Parakaram in 2002, which followed as a response to the terrorist attack on the Indian Parliament on December 13, 2001. In 2006, Pakistan talked about its nuclear policy based on restraint and responsibility and this could be well linked to its efforts to build up its image as a responsible nuclear power in the wake of the India-US nuclear deal, which instilled an acute sense of discrimination and inferiority in Pakistan.

Even though Pakistan's postures regarding use of nuclear weapons have somewhat varied, the ultimate objective behind nuclear weapons has been to guard against India and provide Pakistan the ultimate weapon of security. Even with the fastest growing nuclear arsenal in the world and rapidly multiplying nuclear delivery systems, Pakistan's threat perceptions do not seem to have altered. The threat from India is believed to have gone up in the last decades, partially because of India's rise on the global platform and to a great extent due to Pakistan's continued slide into instability. The strained relationship with the US has also made Pakistan extremely apprehensive of Washington's intentions regarding its nuclear weapons.

Pakistan's projection of irrationality has been articulated extremely rationally and has created enough uncertainty to enhance a level of deterrence. Opting for nuclear weapons in the event of a crisis would have horrendous costs for Pakistan which the defence planners are, no doubt, fully aware of.

Taking a hypothetical scenario wherein Pakistan actually decides to use nuclear weapons in retaliation of an Indian conventional conflict conducted in response to a terrorist attack carried out by state sponsored terrorist groups, its targeting philosophy would run on the principle of first use. The targeting philosophy would not be different from how Pakistan has used its air force with preemptive strikes in the war in 1965 and also in 1971.

Islamabad's objective would be to go in for preemptive strikes to reduce India's retaliatory capability. Secondly, Pakistan has long believed that India finds it difficult to take hard decisions. Hence, it is possible that Pakistan's first strike may target some of India's major cities or at least a threat to target large cities so as to reduce the possibility of a nuclear exchange. A nuclear attack on a major city would generate ample panic globally which would increase the international pressure on India, for not using the nuclear weapons against Pakistan. Given these options, Pakistan would have at least three options in its targeting strategy:

Option I — Target India's nuclear arsenal, in specific, target the delivery systems of the Indian arsenal like the air bases and ballistic missile units.

Option II — Target one or more large Indian cities which would create enough destruction to "paralyse" the Indian command and control system and leadership.

Option III — Try and neutralise the satellite-based communication system to disrupt and possibly black-out the linkages between the National Command Authority, Strategic Forces Command and the launch capabilities.

The fourth obvious option would be to adopt all three options and this may well be the incentive to expand the nuclear arsenal as has been witnessed in recent years.

China has recently constructed and launched a satellite for Pakistan. Officially, this satellite is for earth observation. This takes the China-Pakistan strategic nexus to another level, into space. Hence, it is also reasonable to assume that the Chinese will share the technology for highly accurate ballistic missiles and that for manoeuvrable reentry vehicles. In such case, specially with the Pakistani satellite providing air intelligence, the likelihood of a Pakistani first strike with missiles which are not armed with nuclear weapons will increase and yet achieve a purpose of neutralising a substantive Indian capability to launch its delivery system in case of a nuclear strike.

**Conclusion**

While Pakistan is expected to continue with its position of irrationality, its future nuclear posture could be expected to shape up as follows:

- Pakistan's reliance on nuclear deterrence would intensify with constantly growing asymmetries between India and Pakistan. The increasing differences between the two economies, constant border clashes and Pakistan's politically volatile situation, would increase Pakistan's reliance on nuclear weapons. Pakistan's nuclear weapons in the past two decades have given it the confidence to enhance its capability to wage its covert war in India and have served much larger political and diplomatic objectives for Pakistan. Pakistan is convinced that its posture of irrationality and ambiguity has deterred an Indian aggression more than once and the belief seems to grow with the Indian restraint. But maturity and restraint on the Indian side do not in any way imply lack of Indian capability and resolve to retaliate. Any Pakistani nuclear strike, even if it is with tactical weapons, would face massive Indian retaliation with unacceptable damage for Pakistan. This issue has been discussed in the Indian strategic community. In this context, stressing on the need for stronger Indian signalling, Brig Gurmeet Kanwal holds the view that:

  > India's nuclear signalling must be able to convince the Pakistanis that in response for even one low-yield nuke on Indian forces even within Pakistani territory, India will resort to massive punitive retaliation and will finish Pakistan as a nation-state no matter what damage India might itself suffer.[7]

  This aspect was clearly embodied in the draft nuclear doctrine which was released into the public domain in 1999. (See Appendix I&II)
- Pakistan's nuclear arsenal is expected to continue to expand due to increasing insecurities. Although the number cannot go too far as an increased number would require equivalent increase in the delivery systems, costs and more extensive command and control system, which might not be feasible for Pakistan to acquire. Increase in Pakistan's nuclear arsenal would also have varied implications in terms of increase in concerns regarding the safety and security of

the nuclear weapons even in peace-time. But increase in the arsenal would have a limited role in Pakistan's overall nuclear strategy. It is most likely that they are increasing the arsenal to be able to try and reduce India's nuclear retaliatory capability, by targeting India's nuclear arsenal. Also, it would not be incorrect to assume that Pakistan is increasing the size of its arsenal on the assumption that with China acquiring BMD and other defensive capability, the Indian nuclear arsenal is likely to grow in future. China's role in this aspect becomes critical as China has provided nuclear technology and missiles to Pakistan, constructed their satellite and launched it for them (with more in the future). It is possible that China may also help Pakistan build its capability or at least provide a missile defence cover from Chinese territory.

Hence, looking into these aspects, India must ensure sufficient survivability, even in case of a surprise attack to be able to retaliate and impose unacceptable damage. India is building a BMD, hence, it will be logical to expect that Pakistan will increase its nuclear arsenal. In this context, Air Cmde Jasjit Singh says:

> Under these circumstances, during the times of crisis, a couple of Su-30s armed with nuclear weapons should stay airborne for long periods with aerial refuelling to actually provide the 'assured' counter-strike capability. This should send the message that India's nuclear arsenal cannot be degraded to the level that it cannot impose unacceptable damage, and, hence, make even a minimum deterrence far more credible, independent of a disarming strike.[8]

- Pakistan has been focussed on strengthening the conventional military (especially air power) capabilities to deal with limited war below the nuclear threshold plus greater freedom/confidence for covert war. A strong conventional military base definitely provides confidence to carry on with terrorism, as the nation is strong enough to face Indian retaliation. The US and China have been the main players in building Pakistan's defence capability. While the equipment from the US would slow down, the Chinese supplies are expected to be consistent and increasingly of high technology. The defence modernisation in the last 7-8 years has been centred on the build-up of the Pakistan Air Force and aerial maritime strike capabilities of the Pakistan Navy.

- To sum up, Pakistan's expansion of its nuclear arsenal and conventional military (especially air power and space capabilities) and signals indicating the most likely strategy for its use to be "first use in last resort" indicate raising of the nuclear threshold rather than lowering it, as some people are interpreting. This needs to be carefully watched and constantly evaluated to arrive at a judgement of the extent of strategic space available for any punitive/coercive strategy by India.

**Notes**

1. Hans M. Kristensen and Robert S. Norris, "Pakistan's Nuclear Forces, 2011", *Bulletin of the Atomic Scientists,* 2011 67:91, p.91.
2. For details, see Shalini Chawla, *Pakistan's Military and its Strategy* (New Delhi: Knowledge World, 2009). Also, for details on the Quranic interpretation of war strategies, see Brig S.K. Malik, *Quranic Concept of War* (New Delhi: Himalayan Books, 1986).
3. A mysterious memo was delivered to Adm Mike Mullen, then Chairman of the Joint Chiefs of Staff, asking for US assistance to quell a possible coup in Pakistan. In return, the Pakistan government had offered all manner of concessions to the US, including unfettered access to Pakistan's air space. A wealthy Pakistani-American businessman, Mansoor Ijaz, claimed to have delivered the memo on behalf of Zardari. It is not clear but several reports claimed that the memo was drafted and dictated by Husain Haqqani, then Pakistani Ambassador to the US.
4. "Inconceivable that Osama had no Support System in Pakistan: US", May 3, 2011, *The Hindu.*
5. Ayesha Siddiqa Agha, "Pakistan's Counter-Terrorism Strategy: Separating Friends from Enemies", *The Washington Quarterly,* Winter 2011, pp.149-162.
6. Ibid.
7. Discussion with Brig Gurmeet Kanwal, Director, Centre for Land Warfare Studies, New Delhi, on November 27, 2011.
8. Discussion with Air Cmde Jasjit Singh, Director, Centre for Air Power Studies, New Delhi, January 2012.

**Appendix 1**

# Draft Report of National Security Advisory Board on Indian Nuclear Doctrine

**August 17, 1999**

**Contents**



## Preamble

1.1. The use of nuclear weapons in particular as well as other weapons of mass destruction constitutes the gravest threat to humanity and to peace and stability in the international system. Unlike the other two categories of weapons of mass destruction, biological and chemical weapons which have been outlawed by international treaties, nuclear weapons remain instruments for national and collective security, the possession of which on a selective basis has been sought to be legitimised through permanent extension of the Nuclear. Non-proliferation Treaty (NPT) in May 1995. Nuclear weapon states have asserted that they will continue to rely on nuclear weapons with some of them adopting policies to use them even in a non-nuclear context. These developments amount to virtual abandonment of nuclear disarmament. This is a serious setback to the struggle of the international community to abolish weapons of mass destruction.

1.2. India's primary objective is to achieve economic, political, social, scientific and technological development within a peaceful and democratic framework. This requires an environment of durable peace and insurance against potential risks to peace and stability. It will be India's endeavour to proceed towards this overall objective in cooperation with the global democratic trends and to play a constructive role in advancing the international system toward a just, peaceful and equitable order.

1.3. Autonomy of decision making in the developmental process and in strategic matters is an inalienable democratic right of the Indian people. India will strenuously guard this right in a world where nuclear weapons for a select few are sought to be legitimised for an indefinite future, and where there is growing complexity and frequency in the use of force for political purposes.

1.4. India's security is an integral component of its development process. India continuously aims at promoting an ever-expanding area of peace and stability around it so that developmental priorities can be pursued without disruption.

1.5. However, the very existence of offensive doctrine pertaining to the first use of nuclear weapons and the insistence of some nuclear weapons states on the legitimacy of their use even against non-nuclear weapon countries constitute a threat to peace, stability and

1.6. This document outlines the broad principles for the development, deployment and employment of India's nuclear forces. Details of policy and strategy concerning force structures, deployment and employment of nuclear forces will flow from this framework and will be laid down separately and kept under constant review.

## 2. Objectives

2.1. In the absence of global nuclear disarmament India's strategic interests require effective, credible nuclear deterrence and adequate retaliatory capability should deterrence fail. This is consistent with the UN Charter, which sanctions the right of self-defence.

2.2. The requirements of deterrence should be carefully weighed in the design of Indian nuclear forces and in the strategy to provide for a level of capability consistent with maximum credibility, survivability, effectiveness, safety and security.

2.3. India shall pursue a doctrine of credible minimum nuclear deterrence. In this policy of "retaliation only", the survivability of our arsenal is critical. This is a dynamic concept related to the strategic environment, technological imperatives and the needs of national security. The actual size components, deployment and employment of nuclear forces will be decided in the light of these factors. India's peacetime posture aims at convincing any potential aggressor that :

(a) any threat of use of nuclear weapons against India shall invoke measures to counter the threat: and

(b) any nuclear attack on India and its forces shall result in punitive retaliation with nuclear weapons to inflict damage unacceptable to the aggressor.

2.4. The fundamental purpose of Indian nuclear weapons is to deter the use and threat of use of nuclear weapons by any State or entity against India and its forces. India will not be the first to initiate a nuclear strike, but will respond with punitive retaliation should deterrence fail.

2.5. India will not resort to the use or threat of use of nuclear weapons against States which do not possess nuclear weapons, or are not aligned with nuclear weapon powers.

2.6. Deterrence requires that India maintain:
   (a) Sufficient, survivable and operationally prepared nuclear forces,
   (b) a robust command and control system,
   (c) effective intelligence and early warning capabilities, and
   (d) comprehensive planning and training for operations in line with the strategy, and
   (e) the will to employ nuclear forces and weapons

2.7. Highly effective conventional military capabilities shall be maintained to raise the threshold of outbreak both of conventional military conflict as well as that of threat or use of nuclear weapons.

## 3. Nuclear Forces

3.1. India's nuclear forces will be effective, enduring, diverse, flexible, and responsive to the requirements in accordance with the concept of credible minimum deterrence. These forces will be based on a triad of aircraft, mobile land-based missiles and sea-based assets in keeping with the objectives outlined above.

Survivability of the forces will be enhanced by a combination of multiple redundant systems, mobility, dispersion and deception.

3.2. The doctrine envisages assured capability to shift from peacetime deployment to fully employable forces in the shortest possible time, and the ability to retaliate effectively even in a case of significant degradation by hostile strikes.

## 4. Credibility and Survivability

The following principles are central to India's nuclear deterrent

**4.1. Credibility:** Any adversary must know that India can and will retaliate with sufficient nuclear weapons to inflict destruction and punishment that the aggressor will find unacceptable if nuclear weapons are used against India and its forces.

**4.2. Effectiveness:** The efficacy of India's nuclear deterrent be maximised through synergy among all elements involving reliability, timeliness, accuracy and weight of the attack.

**4.3 Survivability:**

   (i) India's nuclear forces and their command and control shall be organised for very high survivability against surprise attacks and for rapid punitive response. They shall be designed and deployed to ensure survival against a first strike and to endure repetitive attrition attempts with adequate retaliatory capabilities for a punishing strike which would be unacceptable to the aggressor.
   (ii) Procedures for the continuity of nuclear command and control shall ensure a continuing capability to effectively employ nuclear weapons.

## 5. Command and Control

5.1. Nuclear weapons shall be tightly controlled and released for use at the highest political level. the authority to release nuclear weapons for use resides in the person of the Prime Minister of India, or the designated successor(s).

5.2. An effective and survivable command and control system with requisite flexibility and responsiveness shall be in place. An integrated operational plan, or a series of sequential plans, predicated on strategic objectives and a targetting policy shall form part of the system.

5.3. For effective employment the unity of command and control of nuclear forces including dual capable delivery systems shall be ensured.

5.4. The survivability of the nuclear arsenal and effective command, control, communications, computing, intelligence and information (C4I2) systems shall be assured.

5.5. The Indian defence forces shall be in a position to, execute operations in an NBC environment with minimal degradation;

5.6. Space based and other assets shall be created to provide early warning, communications, damage/detonation assessment.

## 6. Security and Safety

**6.1. Security:** Extraordinary precautions shall be taken to ensure that nuclear weapons, their manufacture, transportation and storage are fully guarded against possible theft, loss, sabotage, damage or unauthorised access or use.

**6.2. Safety** is an absolute requirement and tamper proof procedures and systems shall be instituted to ensure that unauthorised or inadvertent activation/use of nuclear weapons does not take place and risks of accident are avoided.

**6.3. Disaster control:** India shall develop an appropriate disaster control system capable of handling the unique requirements of potential incidents involving nuclear weapons and materials;

## 7. Research and Development

7.1. India should step up efforts in research and development to keep up with technological advances in this field.

7.2. While India is committed to maintain the deployment of a deterrent which is both minimum and credible, it will not accept any restraints on building its R&D capability.

## 8. Disarmament and Arms Control

8.1. Global, verifiable and non-discriminatory nuclear disarmament is a national security objective. India shall continue its efforts to achieve the goal of a nuclear weapon-free world at an early date.

8.2. Since no-first use of nuclear weapons is India's basic commitment, every effort shall be made to persuade other States possessing nuclear weapons to join an international treaty banning first use.

8.3. Having provided unqualified negative security assurances, India shall work for internationally binding unconditional negative security assurances by nuclear weapon states to non-nuclear weapon states.

8.4. Nuclear arms control measures shall be sought as part of national security policy to reduce potential threats and to protect our own capability and its effectiveness.

8.5. In view of the very high destructive potential of nuclear weapons, appropriate nuclear risk reduction and confidence building measures shall be sought, negotiated and instituted.

**Appendix II**

## Cabinet Committee on Security Reviews Progress in Operationalizing India's Nuclear Doctrine

The Cabinet Committee on Security (CCS) met today to review the progress in operationaizing of India's nuclear doctrine. The Committee decided that the following information, regarding the nuclear doctrine and operational arrangements governing India's nuclear assets, should be shared with the public.

2. India's nuclear doctrine can be summarized as follows:
   i. Building and maintaining a credible minimum deterrent;
   ii. A posture of "No First Use" nuclear weapons will only be used in retaliation against a nuclear attack on Indian territory or on Indian forces anywhere;
   iii. Nuclear retaliation to a first strike will be massive and designed to inflict unacceptable damage.
   iv. Nuclear retaliatory attacks can only be authorised by the civilian political leadership through the Nuclear Command Authority.
   v. Non-use of nuclear weapons against non-nuclear weapon states;
   vi. However, in the event of a major attack against India, or Indian forces anywhere, by biological or chemical weapons, India will retain the option of retaliating with nuclear weapons;
   vii. A continuance of strict controls on export of nuclear and missile related materials and technologies, participation in the Fissile Material Cutoff Treaty negotiations, and continued observance of the moratorium on nuclear tests.
   viii. Continued commitment to the goal of a nuclear weapon free world, through global, verifiable and non-discriminatory nuclear disarmament.

3. The Nuclear Command Authority comprises a Political Council and an Executive Council. The Political Council is chaired by the Prime Minister. It is the sole body which can authorize the use of nuclear weapons.
4. The Executive Council is chaired by the National Security Advisor. It provides inputs for decision making by the Nuclear Command Authority and executes the directives given to it by the Political Council.
5. The CCS reviewed the existing command and control structures, the state of readiness, the targetting strategy for a retaliatory attack, and operating procedures for various stages of alert and launch. The Committee expressed satisfaction with the overall preparedness. The CCS approved the appointment

---

Prime Minister's Office, Press Release 4th January, 2003

of a Commander-in-Chief, Strategic Forces Command, to manage and administer all Strategic Forces.

6. The CCS also reviewed and approved the arrangements for alternate chains of command for retaliatory nuclear strikes in all eventualities.

# Index